THE ART OF PRODUCING

I believe *The Art of Producing* very rapidly will become the aspiring Producer's reference Bible. It is the most thorough book I've seen that goes in depth to the whole Producer's process of production techniques from start to finish.

Hubert Eaves III – Producer for Luther Vandross,
Madonna, Whitney Houston, Aretha Franklin

The Art of Producing is the first book to standardize a specific production process for creating a successful music project from start to finish. Learn how to develop a step-by-step process for critiquing all of the musical components that go into creating a highly refined production that works for all styles of music. The book provides a well-rounded perspective on everything that goes into producing, including vital information on how to creatively work with bands, groups, and record companies, and offers insight into high-level values and secrets that famous producers have developed through years of trial and error. The book covers detailed production techniques for working with today's latest digital technologies, including virtual recording, virtual instruments, and MIDI tracking. Take these concepts, adapt them to your own personal style, and you will end up with a successful project of the highest attainable quality with the most potential to be become a hit – or just affect people really deeply.

David Gibson has been teaching, engineering, and producing groups in major 24-track studios since 1982 and is the founder of Globe Institute of Recording and Production. He has done recording for James Brown's band, Bobby Whitlock (Derek and the Dominos), The Atlanta Rhythm Sections, Hank Williams Jr.'s band, members of the Doobie Brothers, Lacy J. Dalton, and Herbie Hancock. Gibson is currently the #1 seller of Sound Healing music, also used in hospitals across the U.S., and he is the author of the best-selling book, *The Art of Mixing* (Routledge, 2019).

Maestro B. Curtis, MALS, has produced countless artists, such as Hubert Eaves III (D-Train). He has also produced his own band, Xpression, on the Kalimba Records label with executive producer Maurice White of Earth, Wind and Fire. Curtis has performed with such artists as Peabo Bryson, The O'Jays, George Benson, Phyllis Hyman, and Lenny Williams. He has also played with the Count Basie Orchestra.

THE ART OF PRODUCING

How to Create Great Audio Projects

Second Edition

David Gibson and
Maestro B. Curtis

Routledge
Taylor & Francis Group

NEW YORK AND LONDON

Second edition published 2019
by Routledge
52 Vanderbilt Avenue, New York, NY 10017

and by Routledge
2 Park Square, Milton Park, Abingdon, Oxon, OX14 4RN

Routledge is an imprint of the Taylor & Francis Group, an informa business

First edition published by Artistpro 2004

Library of Congress Cataloging-in-Publication Data
Names: Gibson, David, 1957– author. | Curtis, Maestro B.
Title: The art of producing. How to create great audio projects. / David Gibson and Maestro B. Curtis.
Description: Second edition. | New York, NY : Routledge, 2019. | Includes index.
Identifiers: LCCN 2018031147| ISBN 9780815369394 (hardback : alk. paper) | ISBN 9780815369387 (pbk. : alk. paper) | ISBN 9781351252461 (ebook)
Subjects: LCSH: Sound recordings–Production and direction. | Popular music—Production and direction.
Classification: LCC ML3790 .G539 2019 | DDC 781.49—dc23
LC record available at https://lccn.loc.gov/2018031147

ISBN: 978-0-8153-6939-4 (hbk)
ISBN: 978-0-8153-6938-7 (pbk)
ISBN: 978-1-351-25246-1 (ebk)

Typeset in Zurich Light
by Florence Production Ltd, Stoodleigh, Devon, UK

I know what I want; I just don't know how to do it.

This book is dedicated to all of those who have the music in them and who know what sounds good but don't know how to put the two together.

Contents

Charts and Visuals

Foreword by Hubert Eaves III

As I think back over my 25-plus years of performing, writing, arranging, and/or producing records for such artists as Luther Vandross, D-Train, Stephanie Mills, Madonna, Whitney Houston, Miles Davis, Aretha Franklin, and Roberta Flack, most notably, it has always puzzled me that there was no comprehensive, definitive manual for the art of record production. There was much trial and error, hitting and missing, based on what you could pick up from the ghostly hearsay stories and legends of the proven hit makers who preceded those days. You picked up tips from engineers or assistant engineers, musicians, singers, A&R executives, radio DJs, and so on, as they would speak of their experiences working with whomever the top producers were. Sometimes the producers themselves would give you tips – that is, if you were lucky enough to have a personal relationship with them.

There was a multitude of styles and techniques to influence any aspiring producer as long as you could hear what you wanted and needed for your individual production sound. The problem was [that it was] and still is, to a large degree, very vague as to how to achieve it, other than trial and error and much, much luck.

As a studio musician, I was privy to a wealth of some of those scattered theories and concepts as compared to today, where there is an information overload of fragmented techniques and/or concepts.

Today, you would need to subscribe to a host of magazines to get articles on record production, songwriting, mixing, remixing, engineering, et cetera, as well as surfing the net to buy DVDs, and finding some books that may contain that wealth [of information]. You'll also need to be wealthy to purchase all those different sources of information, and still you may not find all the information that is contained in this all-inclusive producers' bible/manual that has been so intensely put together for those of us always searching for that ever-elusive thing we call "The Hit."

Enter David Gibson and Maestro Curtis. David and Maestro came up with a complete and timely collection of production techniques that is undeniable. Just flipping through the Table of Contents was enough to get me salivating. I was truly impressed and proud to be a friend to Maestro, and now David.

To write the Foreword to *The Art of Producing*, of course I had to read it for myself. I was completely blown away by the wealth of knowledge and information contained herein. Everything you could possibly think of (and including many subjects you may have overlooked) is contained in this one-of-a-kind source for any aspiring or professional hit-maker.

As technology has completely dominated the way we make music today, it's almost impossible to keep up with the advances and products at our command without this book. From working with computer virtual recording, virtual instruments, vocal techniques, virtual effects, and MIDI tracking, to picking a recording studio (even if it is in your own bedroom), to everything in between and more, *The Art of Producing* keeps you well-grounded because the concepts are tried and proven over many years of success, [and there is also] the addition of the new and current visionary techniques.

This book brings light to darkness, or "hit" to miss. I believe they've "hit" it. David and Maestro made it too easy.

—Hubert Eaves III

Acknowledgments

David Gibson's Acknowledgments

There are a wide range of people that helped me along the way to the point where I am writing this book. As with all information, over the years I have simply gathered together a large amount of information from a huge number of contacts and sources – and then there are those divine inspirations, and who knows where they come from?

First, I probably would have never got into this business without the suggestion of my brother, Bill. He was the first to say, "Ever thought about being a recording engineer?" Then, there were my various music instructors and all of my recording instructors, including Bob Beede and John Barsotti. There was also Herbert Zettl, whose book on video aesthetics helped to inspire the structure of this book. Craig Gower was also another inspiring force in learning about working with music. And then there was Chunky Venable, who was kind enough to have the faith in me to run his studio, even though I was such a newbie. Many thanks go to my producer friend, Ken Kraft, with whom I learned many of the techniques within.

There are also those various artists, engineers, and producers who have influenced my values on this long road. Everybody from Pink Floyd to Bob Clearmountain has made a huge impression on my recording and mixing values.

Thanks are in order to the following people: Todd Stock, who has helped with editing and has been a spiritual advisor of sorts; Archer Sully, who has been helping to bring into reality an actual working version of the Virtual Mixer; Silicon Graphics, for their support and for loaning me an Indigo 2 computer; Donna Compton and Patrice Newman, for helping with the rough editing of the book and critiquing of the visuals; Donna, for all her kind and caring support over the years, and to Sharon for her ongoing support; Fred Mueller for some graphics work on the book; extra thanks go to Melissa Lubofsky for her visual inspirations and dedicated work in creating many of the original graphics; and also thanks go to Jon L. Duman for his critique of the section on the music business.

Sincere thanks to Maestro for his incredible amount of experience and musical expertise.

And finally, I would be amiss if I didn't thank all of my students from over the years for the innumerable suggestions and inspirations that they have brought to me.

Maestro Curtis' Acknowledgments

Since the publishing of the first edition of the art of producing, my family has multiplied. So it's only fitting I give acknowledgments and thanks to the following: first and foremost, my source of energy, my wife Nola and my nine children (Zahara, Nile, Isis, Kiki and Phoenix, Fran, Brian, Apollo, and Brandon). I give thanks to the following: Hubert Eaves, Ricardo Scales, Jerome and Cheryl Williams, Suzie Tyner, Pete and Juan Escovedo, Dorothy Morrison Combs, Tyrone Davis, Craig Bicknell, and Darcel Walker. To my mother-in-law Anau Vehikite, my brother-in-law Tomasi Akimeta, to Dwayne Charles, John Honoré, Don Marsh "the Enhancer," Erik LaCoss, John Vitalie, Kevin Myrick, Cathy Davis, Felicia Thibodeaux, London Breed, Chris Borg, Sylvia Sherman, Julie Rulyak Steinberg, the entire Community Music Center program, Masut, Judah, Chris Miller and the entire Haight Ashbury Music family, the Guitar Center, and NARAS the SF Chapter of the Grammys. I include an honorable mention to our Speakeasy family, Nick Olivero, David Gluck, Geoff Libby, and a big thanks to the publishers of our book and the S.O.L. (Spirit of Love) Funkestra, Larry Douglas, Eric Dewalt, Tim Merritt, Winfred Williams, Tyrone Davis, Neil Stallings, and Phil Organo; and lastly, Henry James Larkin. Thank you all, for your love and support over the years. This is dedicated to the memory of Jay'é Richardson (posthumous), who, besides my wife, was my best friend. Maurice White, Joe Sample, and my mother, Lucille (posthumous), all had a major influence on my outlook.

Summary

If you do all the pre-production beforehand, and you get a vibe happening in the control room where everybody is concentrating on every single detail all of the time, you are going to make a huge difference as a producer. Regardless of how much you think you know, just following the procedures in this book is going to help you make a huge difference. By gathering all of the energy together and getting everyone involved, you are on the road to becoming a successful producer. You are now a producer. Repeat after me, "I am a producer." C'mon now, don't roll your eyes. Say it out loud, "I am a producer." You can definitely produce a project now.

Visualize yourself producing . . .
take a deep breath,
and then do it.

Introduction

This book is designed to teach you how to produce a project. It presents simple, step-by-step techniques for how to produce a project, but it also explores the different levels of quality in music, recording, and production. It helps you begin the lifelong process of exploring different value systems in order to create the best projects.

It is interesting how little has been written or taught about what it takes to produce a project and how to go about learning the job, much less obtaining it! It is amazing how many people are unclear about what a producer does in the first place. In fact, there are practically no schools that teach producing. This is understandable because to be successful at the job, you need a wide range of skills from a wide range of disciplines. There is no consensus about how to learn to be a producer, and there is no commonly accepted procedure for producing – until now!

Most producers simply fly by the seat of their pants. Up to this point, most people who have become producers are either born with the necessary talent, or they have been lucky enough to be around a great producer (such as a family member or a coworker). Because of this, there are very few producers in the world. In fact, there is a lack of *good* producers, which explains why producers currently make so much money.

This book is written primarily from the perspective of working on a song or a piece of music that has already been written – otherwise, the book would be about songwriting. As a producer you will often help rewrite parts, but for the most part, the song is already written. However, when it comes to hip-hop production, a whole business has evolved from actually writing the music for the rap artist who has great ideas and lyrics.

The interesting thing about producers is that they make much more money than recording engineers. You can have one hit as a producer and be set for life. You get one hit as an engineer, and you may or may not receive a call for more work.

The goal of this book is to lay out a specific process for producing a project from start to finish. The key for you is to then take this process and adapt it to your own personal style. Ultimately, the goal is to end up with a process and style that works for you and everyone else involved. Throughout the book we provide specific tips for producing a good-quality production, but most important, we spend a large amount of time exploring all the different values of what good quality is.

**Producing is all about developing
high-level values and preferences.**

A clear process has been laid out for how to get there. It still takes work, and you are always learning in order to keep up with changing trends. But that's what makes this industry and this job so interesting. As a producer, you will never be bored.

The book is divided into eight sections:

Part 1: The Producer's Job. This section outlines the three main types of producers – executive, music, and engineering. Most people don't know what the role of a producer is, which is understandable because the role has not been clearly defined. Not only does the producer's job float from role to role, other people (such as the recording engineer and even the band) sometimes function in the producer role. It often leads to confusion.

This section also explains the elusive task of getting the job in the first place. Producer positions are not advertised in the newspaper. We explain the primary ways that people fall into the job and specific techniques for how to get hired.

We then cover the various types of projects to be produced and the skills and knowledge required to do each type of project.

Part 2: The Business Side of Production. This section will help you prepare for and handle a project, and get it done under budget and on time. We explain the functions normally handled by the executive producer. This section includes information on how to obtain financial backing, including various funding sources and how to access them. It also discusses how to budget, organize, and schedule a project, and it includes some forms to help. Although not so common anymore, the section explains the tricky process of working with various types of record companies. Finally, we discuss the new paradigm of self-publishing and promotion.

Part 3: The Virtual Mixer Concept. This section explains the visual framework for displaying imaging, the apparent placement of sounds between the speakers. These visual representations were used extensively in *The Art of Mixing* to explain all the types of mixes that can be created. We use them here to explain the basics of mixing, but we also use them to show how various chord structures and densities of arrangements fit into the overall mix. The truth is that when you add any note or chord to a song, you should think about how to place it in the mix. For example, as you come up with a three-part harmony for the background vocals, you should already be thinking about how it will be placed in the mix from left to right. At the moment of conception, you should be thinking about how much space those parts would take up compared to the rest of the sounds in the mix.

This section also explains the mapping of all the audio parameters of pitch (frequency), panning, and volume as a function of the X, Y, and Z axes in the visual world. This well-known framework can then be used as a tool to help you conceptualize the interaction of all of the dynamics contributing to your overall final production.

Part 4: Music Theory for the Producer. This section explains some of the basic musical concepts in terms that anyone can understand. Even if you already have a degree in music theory, you will gain a different perspective on how basic concepts of music theory can be utilized to create a great production.

This section is organized around the three basic components of sound – time, pitch (frequency), and volume. It covers the basics of understanding how to read and write music, including music notation and values, time signatures, and rhythms; key signatures, intervals, scales, chords, chord progressions; volume dynamics of musicians; and how to read musical charts to enhance and better facilitate the producing process.

Part 5: The Production Process. This section explains all that goes into the two parts of a production – the pre-production process and the role of the producer in the studio. It includes several chapters:

- *Chapter 5: The Pre-Production Process.* This chapter covers what to consider when you are choosing a band (in case you have a choice). It then explains how to introduce yourself to each band member and set up the pre-production meeting. It covers the initial homework you need to do to prepare for the first meeting.

  This chapter then explores some basic people skills that are helpful for making a production meeting go as smoothly as possible.
- *Chapter 6: Dynamic Flows.* In this chapter, we explain the dynamic curves that we experience when listening to a piece of music. We then cover over 20 ways to create more activation and calm, in order to create the ups and downs of these energetic dynamics. Once you understand all the ways to create dynamics in a song, you are able to use these components to create profound intensity or peace.
- *Chapter 7: Structuring and Critiquing the 13 Aspects.* This chapter, the core chapter of the book, covers the details of all of the components that go into creating a great product – concept, intention, melody, rhythm, harmony, lyrics, density of the arrangement, instrumentation, song structure, performance, mix, quality of the equipment and hooks. The chapter explains how to critique each one of the components before the pre-production meeting, and how to present your ideas in the pre-production meeting without stepping on anyone's toes. Most important, it explains how to get the band involved in critiquing each component as well. The key is to be able to pull creative ideas out of the band – to get them involved in the creative process.
- *Chapter 8: Higher-Level Concepts.* This chapter introduces areas that involve a combination of all of the 13 aspects. We cover what it takes to create a certain style of music and how each of the 13 aspects contribute. Although most of us hate to be labeled, the style of music you choose determines the market that you will be breaking into.

  We then discuss considerations that go into deciding the order of the songs on the album. Finally, we discuss how to create a concept album.
- *Chapter 9: Producing in the Studio.* This chapter explains the role of the producer while in the recording session. We discuss how to establish a creative atmosphere in which everyone in the room is focused on critiquing fine details of a performance, and how they fit into the overall quality of the production.
- *Chapter 10: Overall Production Goals.* This chapter covers the ultimate goal of combining all of the components to create a truly great production. We look at the high-level values that major producers utilize in their productions. We also pinpoint what it means to over-produce a project, and how to avoid doing so.

Part 6: Creativity. This section provides a unique exploration into the source of creativity. It is intended to provide you with tools to access the well of creativity that is within all of us. This section consists of the following chapters:

- *Chapter 11: Tapping into the Source: Inspiring Creativity*. This chapter discusses various ideas about the meaning of creativity and where it comes from. It covers many techniques for getting the creative juices going.
- *Chapter 12: Balancing Creativity and Mass Appeal (Industry Trends)*. This chapter discusses the controversial job of balancing pure creativity with making a project with mass appeal. This chapter also reminds you that as a producer, it is important to be aware of industry trends.
- *Chapter 13: Moment-to-Moment Awareness*. This chapter explains the particular level of awareness that is required of the professional producer. It presents two unique approaches to developing a heightened level of sensitivity to the quality of every component that goes into a great production. "David Gets into the Mix" explains a technique for precise listening. It helps you develop a good ear. "Maestro's Composition of the Elements" explains a similar technique that also brings you to an awareness of how all the parts work together when you are writing a song. This chapter also helps you to develop a "multitrack mind" for listening to music. Moment-to-moment awareness is an important skill to develop. It will help you lead everyone in the studio to the same level of discernment. It also gives people a deeper understanding of how to create a higher level of quality.
- *Chapter 14: Songwriting*. This chapter covers basic songwriting techniques. It explores the various types of songs and themes, and provides you with a place to start writing your own music. It also discusses the variety of song formats used in today's popular music.

Part 7: Selecting or Becoming a Producer. This section provides you with the tools for obtaining work as a producer using the techniques you have learned in this book. This section consists of two chapters:

- *Chapter 15: Choosing a Producer*. This chapter explains how to find a producer who fits your style of music and your personality. It explains what to look for in a producer and where to find one. It also gives important techniques for interviewing a potential producer.
- *Chapter 16: Selling Yourself as a Producer*. This chapter explains how to market yourself. It provides suggestions for finding jobs and information about standard practices for compensation.

Part 8: Protecting Your Music. This section provides some basic information on how to protect your music. Copyrights are explained and forms are provided. This section also explains the different types of royalties and outlines the details of all the components of a record deal to help you get involved in the negotiating process.

PART 1

The Producer's Job

Chapter 1: Types of Producers and Productions

Someone produces all records, whether it's an engineer, the band, or a record company staff producer.

During the '40s, '50s, and early '60s, the term *producer* referred to the person who would basically oversee a song or an album. For the most part, the producer's job was simply to keep the band on schedule. However, some producers acted as the eyes and ears for the record company, and sometimes they would also act as cheerleaders, making sure the band was comfortable. Some producers who had developed connections with record executives expanded their jobs to include finding talent and shopping record deals. These individuals called themselves *record producers*.

The job of the producer has now expanded so that he or she is the overseer of quality at every level of the project. But most important, a producer makes sure the overall final product is as perfect as possible. In the old days, the band or recording artist produced without even knowing it, and they never received any credit.

In music production, the goal normally is to produce quality music that will either become a hit or change the world. Today's producers have become the quintessential factors in assuring an album's success. They are often responsible for whether a song, album, or artist will become a hit. Besides overseeing the quality of every component, today's producers often come up with the music. But most importantly they provide an overall energy or vibe to the whole project. It is a cohesiveness that is conceptualized and applied throughout the entire project. To get even more out there, it is an energy or intention that is often carried along on top of the project. It is an invisible essence that people often hear coming through in a song – whether they are aware of it or not. The role a producer plays in a project has become more important than the artist's role, as if it is the producer who makes the artist! What a flip of the script!

Once the producer is a hit, any artist that attaches himself or herself to that producer is likely to be a hit in today's marketplace. In today's world, a successful producer can help make a hit by simply endorsing an artist. This was the case when Kenny Loggins endorsed Michael McDonald. Producers such as Wyclef Jean (Wu-Tang Clan and Lauryn Hill), Kedar Massenburg (D'Angelo and Angie Stone), Bob Ezrin (Pink Floyd), Quincy Jones (everybody), Maurice White (Earth, Wind & Fire, Johnny Mathis, and Barbra Streisand), and Narada Michael Walden (Mariah Carey, Aretha Franklin, and Whitney Houston) have developed such notoriety.

Types of Producers

Producers have evolved into three main types: executive, engineering, and music. However, the job of producer is often misunderstood, and most people really aren't clear about what the job entails. The jobs often overlap and the roles are sometimes quite fuzzy. Producers will often oscillate between the different roles, depending on whichever one is necessary.

Not only do the types of producers overlap, but often the jobs of producer and engineer overlap. Sometimes the engineer or the band might take on some of the functions of producer, and the producer often helps with many aspects that would traditionally be thought of as engineering tasks. For example, producers might help with sampling, sequencing, and looping. Often they get their hands in on a mix. All of these are engineering tasks. In this book I will try to draw the line at the point where producing becomes engineering because there are very good books out there that deal with everything that goes into recording and mixing (including *The Art of Mixing* by David Gibson).

Likewise, engineers commonly function as producers, but no one likes to admit it because then the engineers might ask for more money, royalties, and credits on the album. The truth is that anyone who is making some type of decision regarding the quality of any part or performance in a recording session is really producing.

The producer's job is one of the most important in the recording process. Therefore, it is also the highest-paid job. If Bob Ezrin were to produce the next Pink Floyd album, the record label would probably give him a sizeable advance before he even went into the studio. For the last Pink Floyd album he produced, it is likely that he received seven figures! You know Quincy Jones is set, also. He probably made more than $10,000,000 just from the *Thriller* album alone. It is safe to say that if you get a Top Ten hit as a producer, you are set for life. On the other hand, most engineers don't get any royalties for engineering; they simply get an hourly rate. Eddie Kramer made only about $35 per hour for engineering most of Jimi Hendrix's stuff.

Executive Producer

The executive producer (EP) is responsible for funding, and may also be responsible for organizing a project. He or she might even be responsible for hiring a music or engineering producer for the project.

The individual with the money will often finance a project out of his or her own pocket. The EP is convinced that the artist and product are worthy of investing time and money into, with the sole purpose of reaping the benefits of lucrative record sales and a healthy return on his or her investment. Make no mistake – it is a business! This type of EP may also handle budgets and schedules.

Another type of executive producer is a spokesperson for a record company or a group of investors who fund projects. This type of producer is also responsible for managing budgets and may occasionally schedule the entire process. They generally keep the interests of the record company first and foremost in their minds.

The most common problem arises when whoever is funding the project has no aesthetic sense of the meaning of great art. The worst case is when you have an executive producer who knows little or nothing about music but who wants to start guiding the creative outcome just because he or she is paying for it! Of course, this can become a nightmare, especially when the producer knows only enough to be dangerous! Not only do these types of producers know diddly about music, they often don't have the diplomatic skills that are so important for effective producing.

This is not to say that EPs might not have hearts. In fact, some might even have very refined values when it comes to great art. It is important to get to know your EP as much as possible. If he or she is available, try to develop a close connection. By getting to know your EP and learn his or her values, you often can create a stronger bond that can aid you during rough times. Also, when you learn where this person's values lie, you know where he or she stands if you ever need to put your foot down.

From the other side of the fence, if you are functioning as the executive producer and you have certain wishes, it is better to outline your concerns in writing at the beginning. It can keep you from having to deal with some nasty confrontations later. And the truth is, as will be discussed later, if you have some refined aesthetic sensibility, paying for a project might be an avenue to obtaining your first job as a producer. Many studio owners use studio time as an enticement to obtain the job. All producers have to start somewhere.

Music Producer

Music producers spend most of their time focusing on the music, arrangement, and execution. A music producer arranges the music and often helps to write some of the parts. Sometimes music producers start out as songwriters who have studied music theory and end up with degrees in music or theory; however, some people are born with the knack for arranging. (The latter is not terribly common, though; most music producers have strong backgrounds in music theory.)

In an interview years ago, the band Rush explained that they had just fired their producer because he couldn't keep up with them musically. (Rush's music has always been on the cutting edge when it comes to music theory.) They said they had just hired a real music producer who could help them arrange the parts and who could keep up with them musically.

A music producer can know very little about the mixing board and studio equipment and still be very successful. In fact, we have met a number of major music producers who were not familiar with many of the technical aspects of the equipment. However, a good producer knows to hire a great recording engineer.

> I know everything there is to know about engineering; just ask my engineer.
> —Sonny Limbo (Bertie Higgins, Gladys Knight and the Pips,
> and Alabama)

During a recording session, music producers often are known to spend most of their time with headphones on out in the studio with the band, working out musical ideas.

Besides simply being able to arrange and rearrange music, a music producer needs to know how to work with musicians who have different levels of expertise. Within a single band there are often musicians at different levels. One of the keys to being a great music producer is having the ability to change someone's music and not upset him or her. As you can well imagine, most people are really touchy about their music, especially when someone is rewritiing or reworking it. It is critical to make suggestions with total respect toward what the musicians have already created. You need to get to the point where you can rewrite a band's song and have them totally love it. To do this, you must also be able to gauge whether the musicians will be able to play a particular musical part that you come up with.

Although this book will prepare you for the basics of becoming a music producer, you might consider taking some courses in music theory to add to your knowledge. Many famous producers regret not having gone to school to learn music theory from scratch.

Engineering Producer

The engineering producer's primary job is to monitor the quality of every aspect of the project and make sure it comes out as good as possible. They sometimes also help organize the project. The engineering producer often starts out as a recording engineer, and then moves on to actual producing. Alan Parsons of Pink Floyd's *Dark Side of the Moon* fame started out as an engineer. Likewise, Eddie Kramer (of Hendrix and Led Zeppelin fame) started out as an engineer and later moved into producing. Through the years, Eddie has developed many useful skills as a music producer.

If you have a background in recording engineering, it is relatively simple to move into the position of producer. By the time you finish this book, you will have enough knowledge to produce a group.

Associate Producer

An associate producer is an associate of the executive producer and is there to act as the eyes, ears, and mouth of the EP in his absence. Occasionally, a recording engineer is given this position. He then receives credit (and occasionally royalties) as a producer on the project.

Assistant Producer

An assistant producer is often hired by the producer to do all types of tasks, which could include many of the administrative jobs typically done by an executive producer. The assistant might make progress reports to the record company. But most importantly, the assistant producer is involved enough in the project to take over in an emergency. Occasionally, a recording engineer will be elevated to the level of assistant producer, especially if the producer has to leave for short periods of time.

Overlapping Job Roles

As previously mentioned, the lines of demarcation are not set in stone. The executive producer might get his or her two cents in as a music or engineering producer. A music or engineering producer might kick in some money to help finish a project he or she believes in. Occasionally, an assistant producer might fill in as the primary producer. A producer might end up writing the music, completely rearranging it, and may function as a musician and play a part in the project. For example, Bob Ezrin played many of the keyboard parts for Pink Floyd.

Also, musicians in the band will commonly make production comments and suggestions. Today's producer role seems to have taken on a much broader, yet still intricate, responsibility in ensuring the success of a quality musical presentation.

What It Takes and How to Get the Job

When you finish this book, you will know how to produce – and how to do it quite well. However, to get hired as a producer is a whole other matter – you normally don't see ads in newspapers for producers. Some record companies (majors and independents) do have staff producers who report directly to the head of the Artist and Repetoire (A&R), the department that finds and signs groups, but most record producers freelance and are hired by the artists themselves. Or, an artist might convince the record company to use the new producer.

It is not easy to get someone to trust you enough to hire you and pay you to produce his or her project. There are two basic prerequisites:

1   You must believe that you can produce and say that you are a producer. No one is going to hire you as a producer, if you don't say you are a producer! Say out loud now, "I am a producer," and believe in your heart that you are a producer. To prove it, get some business cards made up with the title "producer" under your name. You are now a producer!

2   To become a producer you must have produced something. The first things a potential client will ask are, "Who have you produced before?" and "Do you have a CD?" Therefore, you need to have produced something. A finished product is your best resume. Remember, talk is not enough; the proof is in the pudding! This book will show you how to get your first production done so you will have something to show off.

As we'll discuss later, the number one way to get a job as a producer is to be recommended by someone you have produced. Another way to get the job is for someone to get sick or die. There have been many examples of engineers getting their break when the producer got called away on another project. The band or producer then might ask him to take over simply because he already knew exactly what was going on. Some people are just in the right place at the right time.

When you get down to it, being a producer requires a wide range of skills, knowledge, and expertise, including:

*   Good listening skills – a good ear, so to speak. A producer must possess the ability to recognize what sounds good and what feels good musically.
*   The ability to organize a large amount of diverse information.
*   The ability to facilitate production and work well with people. Producers must be able to make people comfortable, but they must also be capable of eliciting creativity above and beyond their own expectations. As a recording engineer, it is often important to be as transparent as possible – that is, to not get in the way of the creative process. Although this idea remains crucial to a good production, the professional producer gets to the point where he or she can literally guide the band in the creative process and even help hone the band's style.
*   An understanding of the job of a producer and the producing process.

It is also helpful (but not absolutely necessary) for a producer to have:

- A thorough background as a musician. Today's music producers are likely to be well-crafted (learned) musicians themselves, but this is not absolutely necessary, as mentioned earlier in this chapter.
- A strong background in music theory.
- Knowledge of the techniques of other producers, whether acquired formally or simply by listening to the production of other albums. Many producers mimic or model themselves after producers whom they idolize or look up to in the same way that athletes or artists do. For example, Michael Jordan looked up to Dr. J.; likewise, Mariah Carey learned from Aretha Franklin and aspired to her greatness and level of quality.

The experienced producer is often diverse and is not limited in his or her ability to facilitate and produce any project, regardless of the style of music.

Even after studying this book and learning the procedures for producing a band, you still must embark on the lifelong process of developing detailed values for each aspect of the music.

Types of Projects

Music Production

We're using this category as a catchall for everything except for rap, hip-hop, jazz, and dance music. These other categories are discussed under their own headings in the following sections. For the most part, rock, alternative rock, hard rock, country, reggae, folk, blues, R&B, and neo-soul are all produced using a similar process (with different values applied, of course). This book explains the two-step process in detail: pre-production and the recording session.

Rap and Hip-Hop Production

The big difference in rap and hip-hop production is that you are commonly expected to come up with the musical parts. You aren't simply critiquing them; you are often writing them. If an artist comes to you and says he's got some lyrics, or even if he just wants some music to flow to, you often end up helping to write the song.

On the local hip-hop scene and in the hip-hop world at large, many young artists are looking for producers. They want someone not only to help organize and put the project together, they often want that person to come up with beats and arrange the parts. Everyone and their brother think they can rap, right? Right! Therefore, there is a huge opportunity out there for doing this kind of work in the hip-hop culture.

Dance Music (Electronica)

The big difference with dance music is that most of it is played in clubs, so the orientation is different. In dance music, normally the artist already will have the parts put together in the computer. It is quite similar to a normal project; however, often the person putting the project together is already proficient on the computer. Frequently you end up showing them arrangement tricks and some cool mixing techniques.

Live Recording

Live recording still can include recording in the studio. The term simply means you are no longer using a multitrack. This process, in which the whole band is recorded live, is commonly done for jazz, big band music, and symphonic and classical music. Therefore, you must get a perfect take and mix of the entire band because there is no way to overdub and fix any part, including the vocals.

Commercial Music or Jingles

As with other projects, a Producer of Commercial Music might help write the music or simply oversee the quality of each component. Music for commercials is often much more "in your face," with a much higher level of intensity. Every single moment must be jam-packed with a certain emotion or meaning. Commercials require an extremely strong hook and a much higher level of energy to grab and keep the listener's attention. Essentially, you are trying to get the song's point across in less than one minute.

Audio for Media Production (Video, Film, and Internet)

Producing audio for media requires a whole different set of goals than music production. Often the music supports the visual information. It can be simply background music that is used as filler or to break up monotony. The music might also be part of an actual concert, featured in a movie. Music used to create a mood or anticipation in a film is called a *stinger*. It can enhance or create tension along with the visuals. Music produced for TV involves its own set of traditions that have been developed for how music works with visuals; these are important to learn and there are many books that address these in detail. This also includes video games.

The Producing Process

The producing process normally takes the form of two different procedures – pre-production and production in the recording studio. Here are the steps in more detail:

1 Connecting with the band. This normally means getting to know the band to make sure you are a good match, which often involves seeing them live, going to a few rehearsals, and ultimately getting them to commit to having you produce them.
2 Figuring out what songs you are going to record.
3 Getting the songs on tape for analysis, and getting the lyrics on paper.
4 Doing your homework. This entails analyzing, critiquing, and refining the 13 aspects of a recorded piece of music: Concept, Intention, Melody, Harmony, Rhythm, Lyrics, Song Structure, Density, Instrumentation, Performance, Quality of Equipment and the Recording, Hooks, and Mix.
5 Scheduling the pre-production meetings.
6 Conducting the pre-production meetings. Critiquing the 11 aspects with the band.
7 Scheduling the recording sessions: basic tracks, overdubs, and mixing.
8 Conducting the recording sessions.
9 Mixing.
10 Remixing.

11 Mastering.

12 Completing pressing, packaging, and cover design.

13 Convincing the record comapy not to change anything.

14 Shopping for a record deal.

Many of these aspects are considered the job of the recording engineer. We'll only be covering those aspects that are the main focus of a producer.

PART 2

The Business Side of Production

Chapter 2: Finances, Organization, and Politics

The business side of production is often referred to as *administration*. It involves things such as preparing budgets and schedules, booking recording studios, and clearing musical rights. In certain cases, the producer will even help obtain financing for the project. The executive producer normally will handle the business side of production. However, quite often everyone gets involved – the producer, the engineer, and the band. Again, the job is not well defined (it's a rock-and-roll industry), and the roles often shift. Therefore, regardless of your specific role, it is important that you have a good idea of all that is involved so you can help navigate the processes and educate the rest of the people involved. It is beneficial for everyone involved to have as much information as possible. Occasionally, a producer will hire a production coordinator to handle all of these tasks. As previously mentioned, the producer's basic responsibility is to make sure the project gets done in a timely manner and under budget.

Budgeting, Scheduling, and Organization

Throughout the process of budgeting and scheduling a project, organizational skills are paramount. It is critical that no detail falls through the cracks. One small mistake could mean a loss of a large amount of money.

Creating Budgets

Once the producer receives an allocation of funds from a record company or investor, he must prepare a budget and timeline. If you are seeking funding, you should be aware that investors often require a detailed budget before they invest. If you are dealing with a record company, they will review the budget to make sure it makes sense. If you are dealing with an independent record company, you're often on your own.

You must have a clear financial perspective going into each project. A recording budget can save you lots of money, but most important, it can help you bring a project under budget. A good budget can also help relieve tensions because everyone knows how they will be compensated for their work. As Kashif so aptly described in his widely acclaimed bestseller, *Everything You'd Better Know about the Record Industry*, "A recording budget is, in a sense, like a detailed road map. If planned properly you will arrive at your destination more quickly and with more enjoyment than if you just jump into your car and start driving."

Recording budgets should be detailed and itemized. Your budget should include the number and cost of all producers involved, artists' advances, costs for studio time (including mastering), engineer budget (if it's not included with studio costs), and costs for hired musicians, tape, equipment rentals, travel, and lodging. It is especially helpful to learn and do the budget in a spreadsheet program, such as Microsoft Excel – the primary advantage being that you can easily have all the amounts added up automatically. The chart also helps you to layout the information more clearly. Chart 1 shows a sample production budget.

BUDGET PROJECTIONS					
PROJECT:					
PRODUCER:					
	# OF HOURS	HOURLY RATE	TOTAL	PERSON	DATE PAID
PRODUCER			50000		
ARTIST'S ADVANCES			100000		
STUDIO TIME	*(Get from Schedule)*				
Pre-Production	72	50	3600	Joe	
Recording	410	100	41000	TP	
Mixing	90	100	9000	TP	
Mastering	24	100	2400	Janet	
ENGINEER	*(Get from Schedule)*				
1st	572	35	20020	Tom	
Assistant	572	10	5720	Sam	
MUSICIANS	*(Get from Schedule)*				
Horns	16	200	3200	Jonathan	
Strings	10	200	2000	Horatio	
			0		
			0		
			0		
TRAVEL					
Transportation	500		500		
Hotel	300		300		
Parking	100		100		
SUPPLIES					
Tape	1000		1000		
CD's	50		50		
Harddrives	300		300		
EQUIPMENT RENTAL					
Instruments	500		500		
Mic Preamps	500		500		
Microphones	500		500		
TOTAL			239190		
Signature:					

Chart 1 A Sample Production Budget

To complete the budget, you should have created a schedule of all recording, overdubs, mixing, and mastering (described in the following section, "Scheduling"). However, you don't need exact dates to create the budget. You then take the estimated advance from the record company, make a list of all estimated costs, and make adjustments as necessary.

Often, a record company will allocate an amount for the recording and production, and you get to keep what is left over. So if you get $150,000 from the record company and you spend $80,000 on the recording and $50,000 on the producer, the band gets to split the $20,000 that is left over. However, the record company has the right to recoup all advances, which means you don't get any royalties until all advances have been paid back to the record company. It is estimated that around 95 percent of all bands never make enough money to pay back all the advances.

Scheduling

The producer must also create a complete timeline, and he or she is responsible for scheduling studio time and musicians. A well-organized schedule can really help the sessions to run smoothly because everyone involved knows what is expected of him or her and when. First, the producer should delineate each segment of the project: pre-production, recording, mixing, and mastering.

The key aspect of scheduling involves figuring out which songs and instruments to record first. We'll discuss how to decide the order in Chapter 5, "The Pre-Production Process." Then, you schedule when each of the players is going to play and book the studio time.

Chart 2 (on the following page) shows what a typical schedule might look like.

In this schedule, note that we have included extra time in case it is needed. Under the instruments section, you should note what instruments you will be recording that day. Under the musicians' note, put the musicians' names. You can use the schedule to confirm studio bookings and notification of band members. Also, it is helpful to write down the date you last talked to each musician.

Other Administrative Duties

If you are going to be using a sample from anyone else's album, you must get it cleared (regardless of the length). Frequently, this costs a minimum of $2,000. It is often the producer's job to find the record company and get the clearance. You are also responsible for filing licenses if you are recording other artists' songs or using guest artists from another record company. In addition, you are responsible for obtaining releases from hired musicians and keeping track of all records and receipts for the record company.

Obtaining Financial Backing

Finding a so-called angel investor or financier can be as difficult or easy as you make it. There are very crucial factors to this process that are paramount to successfully acquiring any kind of financial support. These factors include the quality of the concept and product, your belief in the success of the project, your enthusiasm, and your organization and plan of action (in other words, your business plan).

Financial support can come from a range of sources, from the friend or family member who extends a small or moderate loan, to institutions that specialize in lending money. And then

				PRODUCTION SCHEDULE						
PROJECT: FULL PERSPECTIVE						**PRODUCERS: DAVID & MAESTRO**				
EVENT	**PLACE**	**DATE**	**TIME**	**# OF HRS**	**TOTAL**			**STUDIO TIME BOOKED**	**MUSICIANS NOTIFIED**	
PRE-PRODUCTION	**HOURLY RATE = $25/Hour**									
Song 1	Joe's Studio	3/15	10A-4P	6	150			✓		
Song 2	Joe's Studio	3/16	10A-4P	6	150			✓		
Song 3	Joe's Studio	3/17	10A-4P	6	150			✓		
Song 4	Joe's Studio	3/20	10A-4P	6	150			✓		
Song 5	Joe's Studio	3/21	10A-4P	6	150			✓		
Song 6	Joe's Studio	3/22	10A-4P	6	150			✓		
Song 7	Joe's Studio	3/25	10A-4P	6	150			✓		
Song 8	Joe's Studio	3/26	10A-4P	6	150			✓		
Song 9	Joe's Studio	3/27	10A-4P	6	150			✓		
Extra – If needed	Joe's Studio	4/5-4/7	10A-4P	18	45			✓		
				72	1395	**SUB TOTAL**				
RECORDING	**HOURLY RATE = $100/Hour**					**INSTRUMENTS**	**MUSICIANS**			
Basics Songs 1-3	The Place	4/12	10A-8P	10	1000	Drum, Bass, Gtr, Key	JL, DG, KC, RG	✓		
Basics Songs 1-3	The Place	4/13	10A-8P	10	1000	Drum, Bass, Gtr, Key	JL, DG, KC, RG	✓		
Basics Songs 4-6	The Place	4/14	10A-8P	10	1000	Drum, Bass, Gtr, Key	JL, DG, KC, RG	✓		
Basics Songs 4-6	The Place	4/15	10A-8P	10	1000	Drum, Bass, Gtr, Key	JL, DG, KC, RG	✓		
Basics Songs 7-9	The Place	4/16	10A-8P	10	1000	Drum, Bass, Gtr, Key	JL, DG, KC, RG	✓		
Basics Songs 7-9	The Place	4/17	10A-8P	10	1000	Drum, Bass, Gtr, Key	JL, DG, KC, RG	✓		
Overdubs Song 1	The Place	4/20	10A-8P	10	1000	Bass OD, Rhy Gtr	JL, DG	✓		
Overdubs Song 1	The Place	4/21	10A-8P	10	1000	Lead Gtr, Horns	KC, RG	✓		
Overdubs Song 1	The Place	4/22	10A-8P	10	1000	Vocals, BGV's	RG, LT	✓		
Overdubs Song 2	The Place	4/23	10A-8P	10	1000	Bass OD, Rhy Gtr	JL, DG	✓		
Overdubs Song 2	The Place	4/24	10A-8P	10	1000	Lead Gtr, Horns	KC, RG	✓		
Overdubs Song 2	The Place	4/25	10A-8P	10	1000	Vocals, BGV's	RG, LT	✓		
Overdubs Song 3	The Place	4/26	10A-8P	10	1000	Bass OD, Rhy Gtr	JL, DG	✓		
Overdubs Song 3	The Place	4/27	10A-8P	10	1000	Lead Gtr, Horns	KC, RG	✓		
Overdubs Song 3	The Place	4/28	10A-8P	10	1000	Vocals, BGV's	RG, LT	✓		
Overdubs Song 4	The Place	4/29	10A-8P	10	1000	Bass OD, Rhy Gtr	JL, DG	✓		
Overdubs Song 4	The Place	4/30	10A-8P	10	1000	Lead Gtr, Horns	KC, RG	✓		
Overdubs Song 4	The Place	5/1	10A-8P	10	1000	Vocals, BGV's	RG, LT	✓		
Overdubs Song 5	The Place	5/2	10A-8P	10	1000	Bass OD, Rhy Gtr	JL, DG	✓		
Overdubs Song 5	The Place	5/3	10A-8P	10	1000	Lead Gtr, Horns	KC, RG	✓		
Overdubs Song 5	The Place	5/4	10A-8P	10	1000	Vocals, BGV's	RG, LT	✓		
Overdubs Song 6	The Place	5/5	10A-8P	10	1000	Bass OD, Rhy Gtr	JL, DG	✓		
Overdubs Song 6	The Place	5/6	10A-8P	10	1000	Lead Gtr, Horns	KC, RG	✓		
Overdubs Song 6	The Place	5/7	10A-8P	10	1000	Vocals, BGV's	RG, LT	✓		
Overdubs Song 7	The Place	5/8	10A-8P	10	1000	Bass OD, Rhy Gtr	JL, DG	✓		
Overdubs Song 7	The Place	5/9	10A-8P	10	1000	Lead Gtr, Horns	KC, RG	✓		
Overdubs Song 7	The Place	5/10	10A-8P	10	1000	Vocals, BGV's	RG, LT	✓		
Overdubs Song 8	The Place	5/11	10A-8P	10	1000	Bass OD, Rhy Gtr	JL, DG	✓		
Overdubs Song 8	The Place	5/12	10A-8P	10	1000	Lead Gtr, Horns	KC, RG	✓		
Overdubs Song 8	The Place	5/13	10A-8P	10	1000	Vocals, BGV's	RG, LT	✓		
Overdubs Song 9	The Place	5/14	10A-8P	10	1000	Bass OD, Rhy Gtr	JL, DG	✓		
Overdubs Song 9	The Place	5/15	10A-8P	10	1000	Lead Gtr, Horns	KC, RG	✓		
Overdubs Song 9	The Place	5/16	10A-8P	10	1000	Vocals, BGV's	RG, LT	✓		
Extra - If needed	The Place	5/17-5/24	10A-8P	80	8000					
				410	41000	**SUB TOTAL**				
MIXING	**HOURLY RATE = $100/Hour**									
Song 1-2	The Place	5/27	10A-8P	10	1000			✓		
Song 3	The Place	5/28	10A-8P	10	1000			✓		
Song 4-5	The Place	5/29	10A-8P	10	1000			✓		
Song 6	The Place	5/30	10A-8P	10	1000			✓		
Song 7-8	The Place	5/31	10A-8P	10	1000			✓		
Song 9	The Place	6/1	10A-8P	10	1000			✓		
Extra – If needed	The Place	6/2-6/5	10A-8P	30	3000					
				90	9000	**SUB TOTAL**				
MASTERING	**HOURLY RATE = $100/Hour**									
	Da Master	6/5	10A-8P	10	1000			✓		
	Da Master	6/6	10A-8P	10	1000			✓		
Extra – if needed	Da Master	6/7	10A-8P	10	1000			✓		
				30	3000	**SUB TOTAL**				

Chart 2 Typical Recording Production Schedule

there are investors (or "angels") who look for investment opportunities in small businesses or projects for tax-break purposes. Of course, many of these individuals still expect to see some return on their money!

Initial project startup (seed money) might come from several sources, such as family, friends, and even fans who believe in what an artist is doing. Even owners of clubs where artists perform have helped to shape careers by initially backing artists. However, budgeting and saving your own money seems to be the most common way to start. A smart way to begin is to finance a really good three- to four-song demo that can be used to either shop for a record deal or entice someone with cash to invest.

Getting a project financed takes considerable planning. Whether you're financing your own album or demo or someone else is bankrolling your project, you must do some research and formulate a budget for all recording expenditures. This is a smart way for you to map your course as well as save yourself heartaches and money. Your chances of getting a project funded will obviously improve if you present an organized budget in advance.

A large part of getting funded for any type of project requires a development of trust. A budget is good evidence that you know what you are doing and can complete the project within the demonstrated financial constraints. Put together a business proposal listing detailed expenses. There are many good books on how to develop a business plan, and there is also a lot of information and help on the Web. Do a search for "business plans." No one is going to want to invest his or her money without seeing a sound plan of action.

How to find investors and financiers and what they expect might not be as difficult as you might imagine. You would be surprised by how many people with money are looking for small ventures in which to invest money so they can obtain tax write-offs. This is why we stress the importance of networking: It's not what you know (although that helps), but whom you know in this business. So keep your ears to the ground and don't be afraid to make friends. You might also try doing a Google search for "angel investors," just to get an idea of the market.

Investors come in all shapes, sizes, and colors. An investor invests in something that he or she feels is worth his time and money and something or someone he or she believes in. Or, a project that could help a large number of people.

Dealing with Record Companies
Dealing with the major record companies versus an independent label is a whole different ballgame.

Major Record Companies (Majors)
Unless you know someone, getting in the door of a major label can seem almost impossible, and going with a major label often means giving up a great deal of freedom that you would normally enjoy with an independent label (particularly creative freedom, which we will later address in the "Independent Labels (Indies)" section). However, if you're able to get in the door and become successful, the rewards obviously can be huge. That's the big carrot in the sky.

A major record company is a large corporation that has a full stable of artists who are signed to their label. This corporation employs a large staff with executives from top to bottom who run up to 32 departments, including distribution, promotion, marketing, manufacturing, video, artist development, sales, publicity, accounting, legal, and administration, not to mention executive offices that are responsible for maintaining the company's vision and day-to-day operations that help keep the company earning a profit. And make no bones about it, record companies are in the business of making money! As difficult as it may be, if you can view your music as the company does – as a product or commodity – then you will be able to exert more control over the process.

Successful artists realize that it's okay to be creative as well as business-minded. After all, we are talking about your music and your future. Playing music is not a bad way to make money, so wake up and pay attention to the greenery – your greenery. If you don't, the record company will gladly handle it completely on its own!

Most major record companies expect the following:

- Image, with some degree of originality
- A great song
- Singers or vocals
- Talent (although this may not always be the case) that can be developed and exploited
- Reliability
- Sales – that is, a hit record!

For the majors to maximize their function of exploitation and their need to reap profits, sometimes they have to spend millions of dollars, which makes it extremely difficult for the independents to compete. On the other hand, because of the tremendous rosters of major companies, many artists who sign on with a major end up getting lost in the shuffle when priority and attention are given to the label's hottest new act.

As previously mentioned, advances and money spent on the entire process of recording and promoting an artist is recouped by the record company before an artist sees one red cent, and this may take as many as two or three years – if they are recouped at all! Of course, if the record is a huge success, depending on the terms of the contract, that new house could be just around the bend.

Independent Labels (Indies)

It is interesting how few people know that independent record labels have been around for well over 50 years. During the late '40s and '50s, rhythm and blues and country music albums were recorded by indies (who later made the first rock-and-roll recordings), and they changed the recording industry forever! So it is safe to say that the majors would be nothing without the independents. Elvis Presley, who was signed to an indie (Sun Records) out of Memphis, is probably the most recognizable megastar from the early indie movement. Later, Sun sold Elvis' contract to RCA, a major conglomerate. Major record companies then started to realize the benefit of indies.

Chess Records had Muddy Waters and Chuck Berry on their roster. Atlantic Records started as an indie that originally signed Ray Charles, the Drifters, and Ruth Brown, and then later emerged to become a major record company. Probably the best-known independent label to emerge was Motown, under Berry Gordy, Jr. The "Motown Sound" was responsible for acts such as Marvin Gaye, the Four Tops, The Temptations, The Supremes, Mary Wells, Smokey Robinson and the Miracles, Stevie Wonder, and the Jackson Five. In the late 1980s, after

Motown had become a major, Gordy sold it to MCA. Vee-Jay Records had the Four Seasons and the Beatles, but it was eventually taken over by a major. But as you can see, indies helped to reshape the face of music forever.

The benefit of the independent label is
the freedom of artistic expression.

However, this often changes after an artist signs with the conglomerates, which tend to water down the music for so-called mass appeal. Also, when an artist signs with an indie, he or she receives much more attention because indies have a much smaller roster and have to work harder for the success of the artist as well as the label. Most indies use major recording companies and independent distributors, so they don't have a huge overhead. The time and money saved can then be invested back into the label.

As a producer, it might make sense to start your own label. Of course, any time you go into business for yourself there are risks as well as benefits. Starting a record label is no different than opening any other business. Although it is impossible to compete with the majors nationally and internationally, you can manage to become the big fish in the region or even the state. One thing is for sure, if you're a successful indie, the majors will come calling sooner or later.

The obvious advantage of owning your own company is that you get to call your own shots and control the vision of your business. Operating a business requires hard work, dedication, never-ending diligence, and a deep understanding of the music business and its workings. However, as an independent, you can sign fresh new, innovative talent and discover new music.

Self-Publishing and Promotion

These days the most common avenue to make it is to develop your own following. These days publishing and distributing your own music is the most common way to go. Once you have enough sales, the record companies will find you.

Gone are the days of the traditional record companies that were once responsible for signing, developing, recording, producing, promoting, and marketing the artist and their music. Not only were major labels the financiers, they were also responsible for manufacturing and distribution and all costs were known as recoupable. Before the artist could receive one penny, the labels recouped all monies spent as well as their sizeable profits and, in many instances, artists never received one dime. Musicians and artists depended on connections and representatives to shop record deals with the big companies and this was considered to be the big time. Hip-hop/rap artists understood this game and began to empower themselves by removing the middleman and financers and producing, financing, manufacturing, promoting, and distributing their own products. Artists like MC Hammer was said to have sold so many CDs from the trunk of his car that the record labels came knocking on his door to get him to sign a deal.

This trend was becoming the modicum and the new exciting panacea in the hip-hop culture. Rap artists could see immediate residuals on the front and back ends of their creations. This new mindset began to command the attention of artists outside of the hip-hop community like Prince and the entire rock community, as a result record company's value became diminished and infrastructures suffered tremendous decimation. Recording software and equipment became affordable for the everyday layman and the birth of a new movement began to take root in the recording industry: self-empowerment and branding was the new way to a successful music career and potential financial independence.

Home-recording studios, MP3 technology, social media, and video sharing have made the world so much smaller, which has fueled the entrepreneurial spirit. What we have seen from online independent success stories is that good product and consistency is key. Your music is your product and you, along with the images you present to the public at large, are the brand. Great concerts help to introduce and further your music.

Great songs have always been the key to success. However, unlike any other time in the music industry's history, social media has empowered the independent artist. Therefore, good and consistent social media content via websites, Friendster, Myspace (older social networking sites that no longer exist), Facebook, Instagram, YouTube, SnapChat, Podcasts, and others, has allowed the artists to exhibit beautiful photos, even selfies, and live stories of their work, that is, behind the scenes on a music video shoot, recording studio footage, and so on, accompanied by the actual CD.

There is also a growing trend for artists to go with vinyl records, and digital downloads. With the artist now creating his/her own indie label and in some cases partnering with larger independent labels, major record companies are now forced to rethink their format for interacting with the talent. Artists like Katy Perry, one of the biggest icons in pop music today, launched her music career with Myspace, the pre-Facebook social networking site. Perry said: "It is the easiest way to develop your music and develop admirers of your music" (taken from an interview Katy Perry did with Myspace Spain in 2008).

Social media enables artists to develop and reach an international fan base that otherwise would be out of reach and once the artist has established a major following, record labels (those that still exist) get behind those artists and put their machine behind it. For major record labels, such as Capitol and Atlantic, investing in artists that already have their songs recorded and available for purchase, image, and fan base, all of those components, save them money on artist development, production, and recording costs, this in turn allows a successful independent artist a big pay-off and creative control, if and when they choose to sign with a major record label.

Internet Streaming Companies

Again, any way you can get your music out there and develop a following can lead to more sales and visibility. Here is a list of internet streaming sites:

- Spotify
- Pandora
- Google Play Music – All Access
- SONY Music Unlimited
- Mixcloud
- Radio
- iTunes Radio
- Deezer

- Xbox Music
- Soundcloud
- Grooveshark
- Slacker Radio
- Tune-in Radio
- Myspace Music
- Beats Music
- Amazon
- Roundup

Sound Cloud is a music and podcast streaming platform that is free. People can upload original music, find new artists, make their own playlists, to customize their streaming platform designed to allow millions of songs to be heard around the world. Music can be uploaded by artists for free. Music fans can begin to follow different artists, discover lesser known artists and create playlists to enjoy their own custom compilations of artists and music.

Here's a list of the 10 best music streaming apps:

- Amazon
- Apple Music
- Google Play Music
- iHeart Radio
- Pandora
- Slacker Radio
- Sound Cloud
- Spotify
- Tidal
- Tune-in

Regardless of whether you own your own record company or you are seeking to sign with one, it is important to create a network base. Establishing a rapport with others in the industry can lead to many good opportunities and connections. But most important, it provides you with more information to help navigate the industry.

Look for workshops, conferences, and meetings. The music industry has many organizations and support entities that supply information or act as network support systems. Each year many of these businesses and organizations hold seminars, workshops, and awards banquets. Be sure to check out local sections of:

- NARAS – the National Association of Recording Arts and Sciences. They also put on the Grammy Awards
- AES – the Audio Engineering Society
- NAB – the National Association of Broadcasters
- NAMM – National Association of Music Merchants
- Independent Film Society
- Film Festivals

- Music Placement Companies like Taxi
- Music Publishers Directory (Songwriter universe)
- APG, ATLAS Music
- ASCAP
- BMI
- SESAC
- Sound Exchange
- SOCAN PRS for Music.

Digital Distribution Companies

These days there are many distribution channels that you can access on your own. Of course, it is easy to get lost in the mix without advertising dollars, however there are many who have developed their own following and then got picked up by a record company. Here's a list:

- iTunes
- CD Baby
- Orchard Enterprises – this is the company David uses. They distribute to all digital sites.
- Landr
- Tunecor
- Ditto Music
- Blubrry, Studio 21 Podcast, Café Serve Coffee, Expressos – Podcast distribution companies.

Blubrry is an example of a company that provides podcasting hosting, stats, and podcast ads with free WordPress which can be streamed live from one's home. Other hosts such as Studio 21 Podcast, Café Serve Coffee, Espressos and other beverages and who provide ambiance and a studio setting also provide what is widely known as vodcasting, where podcasters actually schedule time to stream worldwide in front of a live studio audience that is viewed on the United Podcast Network, which assures quality production and distribution for podcasters. Podcasts have become such a mainstay that there is now an International Podcast Day.

PART 3

The Virtual Mixer Concept

Chapter 3: Visual Representations of Imaging as a Tool for Production

In this book, we will use visual representations of mixes to explain many of the concepts of music theory used in creating various types of arrangements. This framework gives us a tool to show certain aspects of production without resorting to confusing, extensive explanations. We'll also use this well-known visual framework to explain mixing considerations from a producer's perspective. As a producer, you should think about how every single note you come up with, or rearrange, will be placed in the mix. Therefore, it is important to understand the myriad details that go into creating a great mix. We highly recommend that you pick up *The Art of Mixing*, which covers the subject in fine detail. This chapter explains the basic structure of the visual framework.

Basics of Audio to Visual Mapping – The Space Between the Speakers

We relate to sound in two ways: We feel (and hear) the physical sound waves that come out of the speakers and imagine the apparent placement of sounds between the speakers. The imagined placement is appropriately called *imaging*. The visuals represent imaging, not sound waves. Recording engineers and producers create a wide range of musical dynamics in the world of imaging. Dynamics are defined in this case as anything that creates a change in the listener, whether physically, emotionally, or mentally.

Mapping Volume, Frequency, and Panning Visually

There are three basic parameters of sound, which correspond to the X, Y, and Z visual axes. (See Visual 1.)

Panning as Left to Right

Panning, the left-to-right placement of sounds between the speakers, is naturally shown as a left-to-right placement visually. (See Visual 2.)

Volume as Front to Back

Sounds that are closer to you are louder and distant sounds are softer; therefore, the volume of a sound in the mix can be mapped out as front-to-back placement. (See Visual 3.)

Visual 1
Sound to Visuals: X, Y, Z
Axes

Visual 2
Panning: Left-to-Right
Placement

**Visual 3
Volume: Front-to-Back
Placement**

As you have probably noticed in mixes, some sounds are right out front (normally vocals and lead instruments), while other sounds, such as string instruments and background vocals, are often in the background (thus the term *background vocals*). If you want a sound out front in a mix, the number one thing to do is to raise the fader on the mixing board. Lowering the volume will, of course, put the sound in the background.

Although volume is the number one function of front-to-back placement, you can use other studio equipment to make sounds seem more out front. Boosting an equalizer at any frequency range will normally make it more out front because the overall sound will be louder. Boosting certain frequencies will accentuate the presence of a sound, making it seem even more "in your face." You also can use compressor/limiters to make sounds more out front. They stabilize the sound so it doesn't bounce around so much in volume. When a sound is more stable, your mind can focus on it more clearly, making the sound more present.

Short delays less than 30 milliseconds (ms), which are called *fattening*, will also make a sound more present. (We will discuss more about fattening in the "Size as a Function of Stereo Spread" section later in this chapter.) Also, certain harmonic structures of sounds will stick out more than others. For example, a chainsaw will cut through a mix much more than a flute will. Time-based effects, such as chorus and flanging, and longer delay times tend to make a sound less present simply because it is obscured by a second delayed sound.

Pitch as Up and Down

There is an interesting illusion that occurs with high and low frequencies in the world of imaging – highs are higher and lows are lower. Instruments such as bells, cymbals, and high strings always seem to be much higher between the speakers than instruments such as bass guitars, kick drums, and rap booms. Check it out on your own system. Play a song and listen to where high- and low-frequency sounds seem to sit between the speakers. Height is especially noticeable in a car, so check it out on your car stereo.

**Visual 4
Frequency: Low-to-High
Placement**

There are a number of reasons why this illusion exists. First, low frequencies come through the floor to your feet; high frequencies don't. No matter how much bass you add to a piccolo, it will never rumble the floor. In fact, professional studios are calibrated to exactly how many low frequencies travel along the floor to your feet.

Another reason why highs are high is the fact that your body has a large resonant chamber (the chest cavity) below a smaller resonant chamber (your head). Voice instructors teach you to use these resonant chambers to accentuate different frequency ranges. If you want to bring out the lows, you resonate the stomach.

On a more esoteric level, there are energy centers in the body called *chakras*, which respond to different frequencies. These frequencies are mapped out very specifically from low to high, from the base of the spine to the top of the head.

These energy centers might very well contribute to your perception of highs and lows in the world of imaging. But regardless of why it happens, the truth is that high frequencies do appear

**Visual 5
Frequencies in Us**

**Visual 6
Song with Highs and Lows
Highlighted**

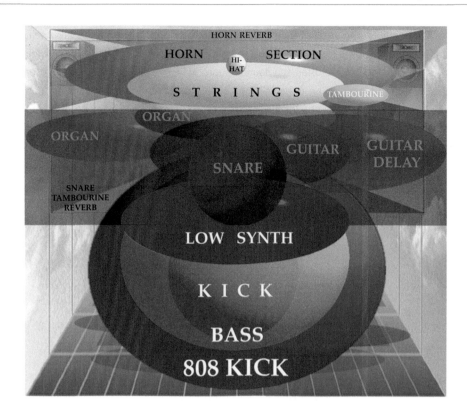

higher between the speakers than low frequencies do. This is also probably why they call high frequencies "high" and low frequencies "low." Therefore, we'll put the high frequencies up high and the low frequencies down low in all our visuals.

You can raise or lower a sound by changing the pitch with harmony processors and aural exciters, or by having a musician play his or her instrument in a higher or lower octave or chord inversion. This becomes important when one sound masks (or hides) another at a particular frequency range. As a producer, it is critical to create an arrangement that spreads sounds evenly over the frequency spectrum so no sound hides another sound – making the mix much cleaner and clearer in the first place.

Because equalization controls the volume of frequencies, you can use an equalizer (EQ) to move a sound up and down at least a little bit, as long as you don't make the instrument sound unnatural in any way. But remember, no matter how much bass you add to a piccolo, you will never get it to rumble the floor, and adding treble to a bass guitar will only raise it up so much.

Defining the Boundaries of the 3D Stereo Field of Imaging

Consider this: The image of a sound never seems to appear farther left than the left speaker or farther right than the right speaker. Right? Right, unless the room is strange. (Sometimes unusual acoustics can make sounds seem to come from odd places in the room.)

Remember, we're not talking about reality here. This is the world of imaging. Because the exact placement is a figment of your imagination, different people see the left and right boundaries differently. Some say that it can't be farther left or right than the speaker itself. Some people with quite active imaginations see sounds as far as a foot or two outside of the speakers. However, most people often see sounds about at least a few inches to the left or right of the speakers (depending on the size of the speakers). Check it out for yourself. Pan a sound all the way to the left or right and listen to see how much farther the image seems to be beyond the speakers.

Here are the left and right boundaries of imaging.

Visual 7
Left and Right Boundaries
of Imaging

Now, take a look at front-to-back boundaries in regard to volume. Just how far do sounds recede into the background as you reduce the volume? How far behind the speakers is a sound before it disappears altogether? Most people seem to imagine sounds to be about few inches to two feet behind the speakers, depending on the size of the speakers. Check out how far back the sound seems to be around various speakers. Normally, sounds are only a short distance behind the speakers.

As previously mentioned, when you turn a sound up, it appears to be more out front in a mix. But how far out front will it go? First, no matter how loud you make the volume of a sound, you can't make it come from behind you. In fact, sounds rarely seem to be more than a short distance in front of the speakers. Most people imagine sounds to be about three inches to a foot in front of the speakers. Again, it depends on the size of the speakers. A loud sound in a boom box will appear only about two inches in front whereas sounds in a huge live sound PA system might appear as far out front as six to ten feet. (Check it out on your own speakers.)

Even Deeper

There is a psychoacoustic phenomenon based on previous experience wherein certain sounds appear to be even farther behind the speakers than the normal imagined limit. For example, if you place the sound of distant thunder between the speakers, it can seem to be miles behind the speakers. The sound of reverb in a large coliseum or a distant echo at the Grand Canyon might also seem to be way behind the speakers. This is a good illusion to remember when you are trying to create unusually expansive depth between the speakers.

Regardless of your perception of the exact limits of imaging from front to back, it is easy to imagine the placement of sounds from front to back, with volume being the main factor that moves a sound. Therefore, the normal stereo field is actually three-dimensional! We'll show the rear boundaries of imaging like this. (The front boundaries aren't shown because they would only get in the way.)

**Visual 8
Front and Back Boundaries
of Imaging**

Finally, what about the upper and lower limits of imaging? As discussed earlier, high frequencies seem to be higher between the speakers than low frequencies. The questions are: How high are high frequencies? And how high do the very highest frequencies we hear seem to be between the speakers? Regardless of the exact limit, sounds never seem to come from the ceiling. Imaging is limited to somewhere around the top of the speakers. Some people say sounds never seem any higher than the speakers themselves, but some say sounds seem to float a few inches above the speakers. Again, the exact limit depends on the size of the speakers and the imagination of the listener.

Now, what about the lower limit? Low frequencies commonly come through the floor to your feet. Therefore, the floor is the lower limit. The upper and lower limits can be shown as in Visual 9.

No matter how far you pan a sound to the left, it will never sound like it is coming from much farther left than the left speaker. (Likewise for the right.) You "see" sounds only a little bit in front of and behind the speakers. And you don't hear sounds higher than the speakers, but they do come through the floor.

Visual 9
Up and Down Boundaries
of Imaging

Therefore, the limits of imaging can be shown like this:

Visual 10
The Only Place Mix Occurs

Speaker size affects your perception of the boundaries of imaging. With a boom box, you normally don't hear sounds more than a couple of inches left or right, in front or back, or above or below the speakers.

Visual 11
Imaging Limits Around a
Boom Box

When you are listening to a huge PA at a large concert, the image might appear as far out front as ten feet, and it might be ten feet behind the speakers. It might easily seem to be as much as six feet farther left and right than the speakers themselves, and it might be much higher and lower than the speakers (see Visual 12).

Visual 12
Imaging Limits Around a
Large PA

This is the space where sounds in a mix occur. In the world of imaging, sounds do not occur anywhere else in the room. Most important, this space is limited, as you can see. Therefore, if you have a 100-piece orchestra between the speakers, it's going to be crowded.

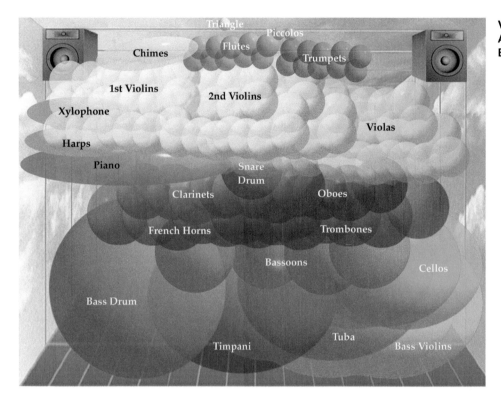

Visual 13
A Large Orchestra Crowded Between Speakers

You can't hear each individual violin in the mix because it is too crowded; you only hear a violin section. In contrast, if you have only three violins, you can hear each one quite clearly.

Visual 14
Three Violins with Plenty of Space In-between

Masking, where one sound hides or obscures another sound, is a major problem in mixing. If you have two sounds in the same place between the speakers, one of the sounds often will be hidden by the other sound. Because the space between the speakers is limited and masking is such a major problem in a mix, the whole issue of mixing becomes one of crowd control.

As you can see, you can move a sound around in the space between the speakers by changing the volume, panning, and pitch (equalization will make small up and down changes).

Visual 15
Movement of Sounds with
Volume, Panning, and EQ

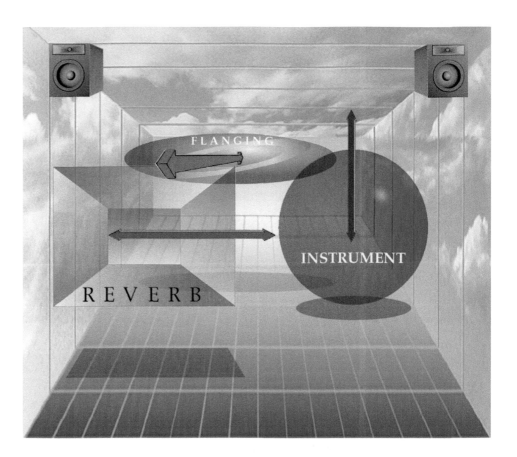

A large part of mixing is simply placing each of the sounds in a different place between the speakers to avoid masking so you can hear each sound clearly. However, as we'll discuss later, there is a bit more to it. Sometimes you just might want sounds to overlap, creating a fuller type of mix (instead of placing sounds apart from each other). You might want some sounds to overlap and others to be separate in order to highlight them. There is a huge number of possibilities.

You might think that if masking is such a big issue, then it is important to know how much space a sound takes up in this limited world of imaging. In fact, not only do different sounds take up different amounts of space, equalization and effects can make a huge difference in how much space a sound uses.

How Much Space Does It Use?

Just how big is each sound in this world of imaging? With limited space between the speakers, you need to know the size of each member of the crowd so you can deal with the problem of masking. The more space a sound takes up, the more it will hide other sounds in the mix.

Visual 16
Solar Eclipse: Natural Masking

More Space

At first thought you might think that you could make the space bigger by moving the speakers apart. The only problem is that the sounds become proportionately larger in size, so you end up with no more space than you started with. On the other hand, a 3D sound processor allows you to place sounds outside of the space between the speakers; it expands the space between the speakers. Surround sound (5.1 or any type of multichannel mixdown system) also expands the three-dimensional space between the speakers to include the entire room.

Visual 17
Surround Sound Mix

Size as a Function of Frequency Range

First, bass sounds seem to take up more space in the mix than high-frequency sounds. Place three bass guitars in a mix, and it will be muddy.

**Visual 18
Mud City**

Bass sounds, being bigger, mask other sounds more. However, place ten bells in a mix, and you can still discern each bell distinctly – even if they are all playing at the same time. High frequencies take up less space in the world of imaging.

**Visual 19
Ten Bells Playing at the
Same Time**

Therefore, the visuals representing high-frequency sounds are smaller and placed higher than the low-frequency instruments, which are represented by larger images and placed lower between the speakers.

Size as a Function of Volume

The louder a sound is in the mix, the more it will mask other sounds. Therefore, louder sounds are larger. A guitar that is extremely loud will tend to mask the other sounds a lot more than if it were soft. A bass guitar, already large, will hide other sounds even more when it is loud.

**Visual 20
A Loud Bass Guitar Masking the Rest of the Mix**

Non-Directional Lows

Technically, it is very difficult to tell exactly where low frequencies (those below 400 Hz) are coming from. Low frequencies are extremely difficult to localize between the speakers. Therefore, a more realistic visualization would have the low-frequency spheres less defined – spreading out to cover the entire lower portion of the visual – creating even more masking. However, to be able to show the specific volume, panning, and EQ of bass, we will continue to use large, defined spheres.

Size as a Function of Stereo Spread

When you have a delay longer than 30 ms (remember, 1,000 ms = 1 second), you hear two sounds. (See Visual 21.)

An unusual effect occurs when you put a delay on a sound less than around 30 ms. Because your ears (and brain) are not quick enough hear the difference between delay times this fast, you only hear one fatter sound instead of an echo. This effect is commonly called *fattening*. When you place the original signal in the left speaker and the short delay in the right speaker, the effect is such that it "stretches" the sound between the speakers. (See Visual 22.)

It doesn't put the sound in a room (as reverb does); it simply makes the sound omnipresent between the speakers.

Visual 21
A Delay Longer than
30 Milliseconds

Visual 22
Fattening: <30 ms Delay Time

Placing two microphones on one sound can create a similar effect. Because sound is so slow (around 770 mph), you get about 1 ms of delay time per foot. Therefore, when the two mics are panned left and right between the speakers, you will hear a short delay that also creates a stereo sound.

Visual 23
Close to 1 ms Delay Time
Per Foot

Additionally, sounds in synthesizers are commonly spread in stereo with these same short delay times.

Visual 24
Fattening panned 11:00–1:00

Just as you can use volume, panning, and EQ to place and move spheres, you also have control over the placement of the oblong sphere, or "line" of sound created by fattening. You can place the line anywhere from left to right by panning the original signal and the delayed signal to a variety of positions. The wider the stereo spread, the more space the sound takes up and the more masking it causes.

Visual 25
Fattening panned 10:00–2:00

You also can bring this line of sound up front by turning the volume up,

Visual 26
Loud Fattening Right Up Front

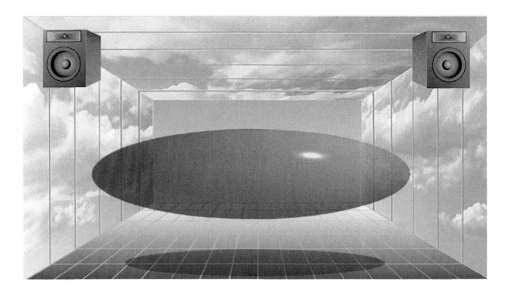

or place it in the background by turning the volume down.

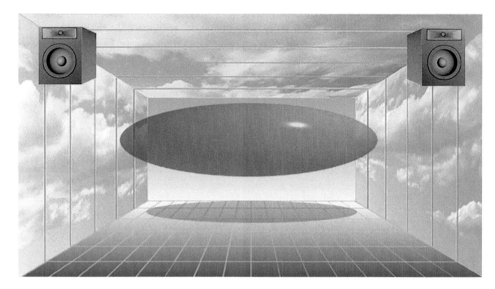

Visual 27
Low Volume Fattening in the Background

You also can move it up or down a little bit with more treble or bass EQ.

Visual 28
Fattening with High-Frequency EQ Boost

Visual 29
Fattening with Low-Frequency EQ Boost

Size as a Function of Reverb

Placing reverb in a mix is like placing the sound of a room in the space between the speakers. A room, being three-dimensional, is shown as a 3D, see-through cube between the speakers. Reverb is actually made up of hundreds of delays. Therefore, it occupies a huge amount of space when it is panned in stereo. It is as if you are placing hundreds of copies of the sound at hundreds of different places between the speakers. This is why reverb causes so much masking!

Visual 30
Stereo Reverb on Sound

Just as you can place and move around spheres and lines of sounds in a mix, you also have control over the placement and movement of reverb with panning, volume, and EQ. You can place reverb anywhere from left to right by panning the two stereo outputs of the reverb in a variety of positions. The wider the stereo spread, the more space reverb takes up and the more masking it causes. (See Visuals 31 and 32.)

Visual 31
Reverb Panned 11:00–1:00

Visual 32
Reverb panned 10:00–2:00

When you turn the volume level of the reverb up (which is normally done by turning up the auxiliary send on the sound going to the reverb), it comes out front in the mix.

Visual 33
Loud Reverb

With EQ, you can raise or lower the placement of the reverb a little, which makes the reverb smaller (more trebly) or larger (bassier). (See Visuals 34 and 35.)

Visual 34
Reverb with High-Frequency
EQ Boost

**Visual 35
Reverb with Low-Frequency
EQ Boost**

You can place these three basic sound images – spheres, lines, and rooms – within the three-dimensional stereo field between the speakers to create every structure of mix in the world.

Spheres represent sounds, oblong spheres represent fattening, and translucent cubes of light represent reverb. All other effects, including different delay times, flanging, chorusing, phasing, parameters of reverb, and other effects, are variations of these three images and will be described in detail in David Gibson's book, *The Art of Mixing*. With these various sound images, you can create a wide range of mix styles appropriate for various music and song styles.

For example, you can create even versus uneven volumes (see Visuals 36 and 37),

or balanced versus unbalanced mixes (see Visuals 38 and 39),

or natural versus interesting EQ (see Visuals 40 and 41),

and sparse versus full (wall-of-sound) mixes with effects (see Visuals 42 and 43).

Visual 36
Even Volumes

Visual 37
Uneven Volumes

**Visual 38
Symmetrical (Balanced) Mix**

**Visual 39
Asymmetrical (Lopsided) Mix**

Visual 40
Natural EQ

Visual 41
Interesting EQ

Visual 42
Sparse Mix

Visual 43
Full (Wall-of-Sound) Mix

This limited space between the speakers where imaging occurs is the *stage* or *pallet*, where you can create different structures of mixes. An engineer must be adept at coming up with any structures and patterns that can be conceived. Each one of these structures creates different feelings or dynamics. Just as a musician needs to explore and become thoroughly familiar with all the possibilities of his or her instrument, a producer must be aware of all possible dynamics that the equipment can create.

The art of mixing is the *creative* placement and movement of these sound images. The mix can be made to fit the song so the mix becomes transparent or invisible. Or, the mix can be used to create musical dynamics of its own. It can be a tool to enhance and highlight, or it can create tension or chaos. A great engineer or producer uses the mix to push the limits of what has already been done.

The art of mixing is also the *appropriate* placement and movement of these sound images. Once you know how to create any style of mix, the key is to create mixes that fit the music and song in some creative way.

As a producer, one of the key aspects that affects the clutter in this 3D space is the number of instruments or sounds playing at any one time, and the height which is a function of the average of the pitch of the sound.

The arrangement and choice of instrument sounds is the number one thing that makes a good mix! With a good arrangement the mixing engineer has very little to do other than create really fine aspects of magic in the mix.

The Producer as Sculptor

An engineer or producer has the same range of control as a sculptor, artist, architect, or Feng Shui consultant: All are working in 3D. The sculptor deals with shaping the images in a three-dimensional space. In photography and painting, the artist deals with depth perception, the way colors interact, and composition of all the components. In construction, the architect deals with creating a space that feels right. He begins with a strong foundation (basic rhythm tracks – drums, bass and guitar, or keys). In Feng Shui, the consultant deals with placement of elements in a 3D space. Here we are dealing with the Feng Shui of mixing.

You now have a framework that you can use to show how musical dynamics work with mix dynamics for a great overall production.

PART 4

Music Theory for Recording Engineers and Producers

Chapter 4: Music Theory Primer

This chapter serves as a practical reference guide essential to understanding the basic fundamentals of music theory, which are pertinent to the performing, recording, and production processes. This basic information in music theory will not only provide you with insight into the creative process, but it also will allow you to be properly prepared for the sometimes bumpy musical journey, whether you're an engineer, producer, or executive. The more proficient you become, the more you increase your value in today's progressive recording industry marketplace.

We will provide you with the rudiments of music theory in the most basic terms, emphasizing the most significant and widely used terminology and applications common to musicians working in the industry today. Although the world of music theory is as expansive as life itself, we've tried to give you the information that is required for production. No fluff.

It is important to understand the basics of reading and writing music so you can keep up with a band who is discussing their music using musical terms. It is also helpful for you to be able to follow along when someone brings in the music written out. The following information is extremely useful in understanding how to read the chord charts that are used by musicians in the studio. It is also helpful to know, so you can map out and come up with new musical parts in the pre-production process.

The ability to read charts or music will give you more insight into the mind and processes of a songwriter. It is enormously rewarding, too. With this information you will not only be able to follow the music, you will be able to help identify any mistakes or deviations.

From an engineering perspective, understanding how to read chord charts helps in the creation of a map of the mix throughout the song. This will help you keep track of where any automation information is entered. When you are using a sequencer or digital audio workstation (DAW) with a click track, you will be dealing with software that is displaying the recorded information in standard measures anyway.

The more you work on music using a computer, the more you will encounter components of music theory. This basic theory will help you to navigate through various musical styles, which will allow you to build a much more solid background. Although some musicians and music thinkers have had success without rudimentary music instruction, fundamental knowledge of music theory has helped people to be more skilled, equipped, and open to exploring all musical ideas and possibilities. This chapter is designed to encourage and inspire you to build, explore, and expand your own musical creativity. It also provides you with a broader outlook on music. Having both the creative inspiration and the knowledge of how music works can make you a complete musical force!

Music is displayed on a five-line staff that is divided into measures (also called bars), with notes and rest signs of varying lengths. There are notations at the beginning of each section that display the key of the section (showing all flats and sharps) and the time signature, which shows how many quarter notes there are per measure.

Visual 44
Staff with Time Signature, Key
Signature, Notes, and Rests

As discussed in the previous chapter, there are three basic components to sound – frequency (pitch), time, and volume. If you look at each component of sheet music from this perspective, you can see how each note and its performance affect the overall structure of the mix for the final production. We'll begin with time in sheet music.

Time

This section explains how to read rhythms and will provide you with an understanding of the distinction between beats per minute (bpm), or *tempo*, and time signatures.

If you have a tempo of 60 beats per minute, you have one beat every second. In a graphic editor in sequencer software, four beats are displayed as four long lines, as shown here:

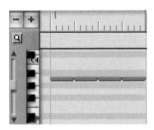

Visual 45
Measure with Four Beats on a Sequencer

In sheet music, it is displayed as shown here:

Visual 46
Measure with Four Beats on a Staff

A beat is normally represented by a quarter note. Therefore, at 60 bpm this whole measure would be four seconds long with one beat every second.

Note Values

Understanding note values allows you to count musically and keep up with where you are in the song. It is especially helpful to be able to count along so you can figure out where to punch-in during an overdub (for example, when fixing one line on a particular track by recording over just that one line).

Note values indicate how long the note is to be played or held.

Visual 47
Digital Performer (DP) Staff with Four Measures Showing One Whole Note, Two Half Notes, Four Quarter Notes, and Eight Eighth Notes

On a sequencer these note values would look like what you see here.

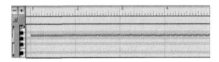

Visual 48
DP graphic editor with four measures showing one whole note, two half notes, four quarter notes, and eight eighth notes

The graph in here shows the relationship of all note values.

NOTE VALUE PYRAMID

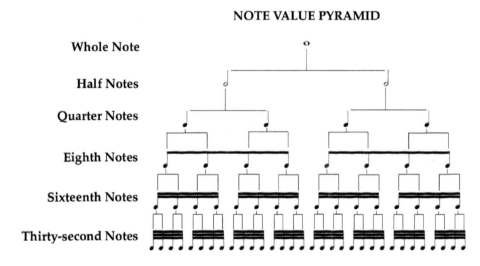

Whole Note

Half Notes

Quarter Notes

Eighth Notes

Sixteenth Notes

Thirty-second Notes

Visual 49
Note Value Pyramid

On a typical sequencer there are 960 ticks per quarter note (although some sequencers only have 480 ticks per quarter note). This means that an eighth note is 480 ticks, a sixteenth note is 240 ticks, and a thirty-second note is 120 ticks. If your sequencer has a resolution of 480 ticks per quarter note, then just cut the number of ticks for each note value in half.

You can see that a graphic editor shows more details of minute timing differences. This allows you to really tweak the feel of a rhythm by having notes played barely before or after the beat. Sheet music is actually quantized (moved to the nearest beat), so is not nearly as detailed as graphic editors, which have such a fine resolution of 960 ticks per quarter note. However, compared to graphic editing, sheet music is much easier to read and play along with.

A *dotted note* is note with a dot. The dot increases the note's duration by half again its value. For example, a dotted half note would represent a half note and a quarter note combined in duration.

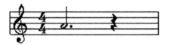

Visual 50
A Dotted Half Note on a Staff

On the sequencer you can see precisely how long the note lasts. Here we have zoomed in, dividing one measure into four beats.

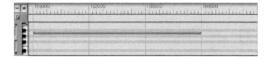

Visual 51
A Dotted Half Note on a Sequencer

A *rest* is the silence or pause in the music between notes. The rest's silence value is the equivalent of actual note values.

Visual 52
Rests on a Staff

In a sequencer, these same rests would look like what you see here:

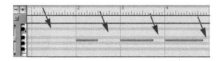

Visual 53a
Rests in a Sequencer

Circle Rhythm

Here is a unique way to visualize rhythms using a circle. You might find this visualization when working on rhythm patterns. You can also even tap on the circle.

Here is a four beat circle which would show one measure.

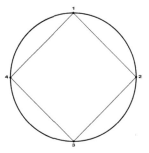

Visual 53b
4 Beat Circle

Here's an 8 Beat Circle.

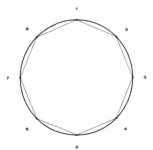

Visual 53c
8 Beat Circle

Here's an 16 Beat Circle, broken up into 4 measures.

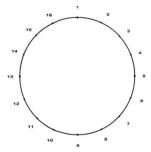

Visual 53d
16 Beat Circle

Time Signatures

The *time signature* refers to numbers that look like fractions, which appear at the beginning of the measure. As you can see below the fraction is 4/4. The top number represents how many counts or beats are in the measure. The bottom number tells you what note value (whole, half, quarter, eighth, or sixteenth) is used to represent one beat. For example, if the bottom number were 8, an eighth note would be one beat.

Different time signatures are commonly used in specific styles of music. For example, 4/4 is common in pop music, rock-and-roll, R&B, hip-hop, and especially dance music. A waltz is 3/4 time. Some classical music is also commonly in 3/4 time. Country and gospel music often use 6/8 time. Jazz and classical commonly utilize a wide range of time signatures and even combine more than one time signature in a single song.

A *polyrhythm* occurs when you have two or more different rhythmical patterns and/or two time signatures occurring at the same time in a song. One example is when the hi-hat is playing 6/8 and the kick is playing in 4/4, or a completely different instrument might be playing a different time signature feel. This is what that would look like on a circle template.

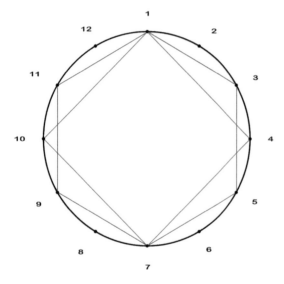

Visual 53e
Circle Template Polyrhythms

Tempo

The *tempo* refers to how fast or slow a piece of music is to be played. It is calculated in beats per minute (bpm), but it is often written in European language above the staff.

European Language	Approximate BPM	Description
Largo	40–65	Funeral or Sound Healing
Andante	50–65	Normal slow walk
Adagio	65–75	Slow
Mezzo	75–110	Moderate

Allegro	110–120	Moving quickly
Vivace	120–200	Very fast
	120	Breakbeats
	130	House
	140–150	Trance
	170	Jungle (Breakbeats)
	180	Hardcore (House)
	200	Gabber
	40–80	Sound Healing
	50–80	Ballad
	90–115	Hip-hop
	90–120	R&B
	100–120	Standard rock-and-roll
	140 and up	Bluegrass

Ritardando (commonly called a *retard*) is where the tempo gradually slows down, normally at the end of a song or section.

Many people who write music often come up with their own descriptions to indicate the tempo, such as "Play moderately fast." When you know the people you are working with, whatever works is fine.

Rhythmical Structures

You can use counterpoint and syncopation to spice up a rhythm part. *Counterpoint* is when an instrument is playing a fraction or multiple of a note value compared to another instrument. For example, suppose you have strings playing a whole note, horns playing half notes, and a guitar playing quarter notes. The key to good counterpoint is having instruments playing their notes on different beats so they do not step on each other. This is a basic technique of James Brown's band, and it is also common in electronica.

Syncopation

Syncopation is when the accent in a musical phrase or passage falls on the weak beat instead of the strong beat. When you play an instrument on an offbeat of the main rhythm, it makes the rhythm bouncier. If you are counting, "One and two and three and four and. . ." (where the "and" is the offbeat), then the accent would fall on the "and." Each of these techniques provides for more complexity and can add more emotion into the rhythmical content. However, they can get in the way if they are too busy. Most funk, R&B, hip-hop, gospel and dance music use rhythmical syncopation to create the dips for dance.

You can use *rhythmic accents* to create a natural rhythm or to create something more interesting. For example, if you are setting the accent on beat 1, it is similar to a marching band. In most rock-and-roll, the accent is on 2 and 4, on the snare drum. When the accent is on the 2 and 4 this is called the backbeat. In jazz, gospel and even some classical music, accents falling on the up beat are just as important. Using other beats as the accent can create some very unique excitement. For example, the song "Billy Jean," by Michael Jackson, had the accents on 1 and 4.

Here is a clapping exercise. Try clapping this exercise with a steady beat, clap louder where the accents are (^). You are now clapping in syncopation.

A *grace note* is a passing note in the rhythm that is de-emphasized by playing softer instead of louder. They help to make a rhythm more complex and interesting. Grace notes often are used in dance music. Almost every groove James Brown has come up with has grace notes that move the rhythm along.

Pickup

A *pickup*, or *lead-in*, is musical part that begins just before a section of the song or a lead part. A pickup can smoothly bring in a melody, or it can be used to introduce a rhythm.

Pitch

In sheet music, *pitch* is shown as a function of the height on a staff, which consists of five lines and four spaces, each representing a musical note, sound, or tone corresponding to the musical notes on the piano.

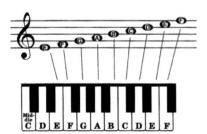

Visual 54
Notes on a Staff Above Notes on a Piano

On a sequencer the same notes would look like what you see here. (Note the piano keys on the left.)

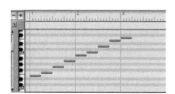

Visual 55
Notes in a Sequencer

Musical notes follow the alphabet from A to G (ABCDEFG) and then repeat. When displaying piano music, the scale normally goes from C to B (CDEFGAB) and then repeats at higher or lower octaves. The full eight notes from C to C are referred to as an *octave*. Each C is an octave above or below the next one.

There are also notes between most of the notes. If you go a half-step down, it's called a *flat*; a half-step up is called a *sharp*. Therefore, there are actually a total of 12 half-step notes before the octave repeats. On the staff, symbols are added to display whether notes are to be played as a flat or sharp.

Visual 56
Notes with Flat and Sharp Signs on a Staff

In a sequencer, the same notes would look like what you see here:

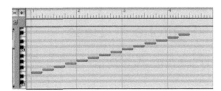

Visual 57
12 Notes in a Sequencer

A *staff* may refer to higher or lower notes on an instrument. *Clef signs* are symbols that are written at the beginning of the staff and are used to indicate whether notes are to be played higher or lower. Higher notes are displayed by the treble clef (G clef) from tenor to soprano and beyond.

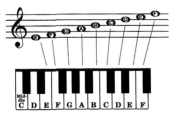

Visual 58
Treble Clef

The treble clef is called a *G clef* because it is a stylized G whose base encircles the G line on the staff.

The *bass clef* (F clef) is used for notes that are pitched in the lower register or range of an instrument.

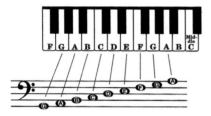

Visual 59
Bass Clef

Chart 3 shows the typical ranges from soprano to bass with the corresponding frequencies in Hertz (Hz).

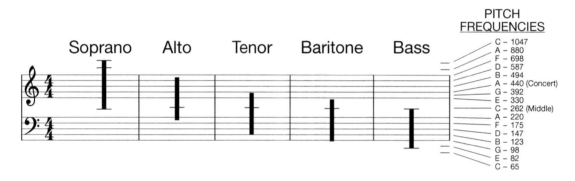

Chart 3 Soprano to Bass and Frequencies

As you might have noticed, there are only nine positions on a staff. Notes will often extend above and below a staff with the use of *ledger lines*.

Visual 60
Ledger Lines

When both staffs are combined, the resulting staff is called a *grand staff*. It is used for piano music, where the left and right hand are playing different things.

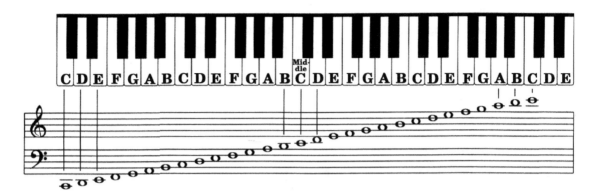

Visual 61
Grand Staff

The single line between the two staffs indicates middle C.

As you might have noticed, the grand staff really only shows about four octaves. Most pianos actually span around seven octaves. Therefore, piano players are given symbols to indicate that a particular note is to be played higher or lower than the staff shows. The symbol is "8va" with a line pointing up or down.

> A really good piano arrangement can almost cover a full band arrangement, depending on song type or style of music (R&B, country, neo-soul, and most popular contemporary music), whereas symphonic music requires an elaborate conductor's score that can include as many as 32 separate parts for a 64-piece orchestra with the piano score as just one of the parts.

As a producer, you should know what key a song is in. Different keys have more or fewer sharps and flats. If you play a note that is not included in the key it feels funny – most people would say it is wrong. The whole song gravitates to the main note or home base, which is the key. This home note in a key is called the *tonic*.

The sharps or flats located at the beginning of the music on the staff following the time signature indicate the key in which the song is to be played. It is called the *key signature*. If there are no sharps or flats, in most cases the song is in the key of C.

Visual 62
Key Signature

To undo a key signature at the beginning of the song, a natural symbol is placed in front of a particular note. It cancels the key signature only for that one measure.

Musical Intervals

More important than any single notes is the relationship of notes to one another. The distance between each note is called the *musical interval*. For example, the interval distance between C sharp and D is a half-step The relationship between C and D is a whole step. A key part of music is the different feelings and emotions that different intervals invoke in us. Though it is subjective so people get different feelings, there is a commonality to what we experience for each interval. The intervals are shown here:

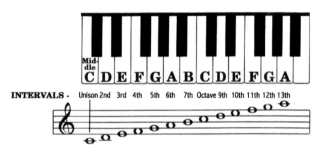

Visual 63
Piano Intervals (Unison, 2nd, 3rd, 4th, 5th, 6th, 7th, Octave, 9th, 10th, 11th, 12th, 13th)

The feelings associated with intervals are shown in the following table (see pp. 71–73).

Scales

A *scale* is a specific group of notes that are commonly played when someone plays a solo. A *major scale* consists of eight notes and is the basis of how we relate to chords, melodies, and other scales. The major scale can be considered the ground zero or starting point by which to identify notes. One cannot make reference to minor, diminished, augmented, or any alteration without first fully understanding and feeling the major scale. It is sometimes referred to as the Ionian scale. Naturally, you would think that a scale with the notes CDEFGABC (the major scale in C) has a whole step between each note. Instead, it has the step pattern of WWHWWWH (where W is a whole step and H is a half-step).

For every major scale there is a relative *minor scale* (which follows the pattern of the natural minor scale shown in the following table). You could have someone play in the relative minor to keep the music interesting. It is another option for a lead part that provides a completely different feel.

ARCHETYPAL RELATIONSHIPS OF MUSICAL INTERVALS		
Interval	**Description**	**Musical Examples**
Unison (1:1)	Two identical pitches sounding together associated with primal cosmic union which represents serenity and perfect peace. Very comfortable interval. There is a sameness about the sounds, as it is in complete harmony with the fundamental sound.	"Row, row, row your boat" "Happy Birthday"
Octave (1:2 or 2:1)	Generated by two sounds where one is two times the frequency of the other. This interval is grounding, meditative, calming, also called the yin and yang interval. One tone is now representing the male and the higher note being female or feminine. Final resolution. Reaching for the higher self. Foundation for stability. Linking past to the present memory. Transformational. Brings resolution and ascendance in the spiral of evolution in sound. Home again, at a higher or lower level. Dimension and interrelation. Bright, unity, joyful, high. Higher significance of something. Transition, transmutation, death, rebirth, finishing and beginning.	"Somewhere over the Rainbow" "Let it Snow"
Perfect Fifth (3:2)	This interval is comfortable. It is the first overtone, evoking feelings of home, sturdiness, and completeness. A very nice interval that is pretty far off from the tonic. Awe-inspiring, desolate. The 5th opens channels that activate power and motion. This is also a pivot interval which can bring you back to the one or lead to the octave creating a full circle. Creation, Actualizaton. Bright, contemplative, full of light, heavenly or divine (when gentle). It is the opposite of the 4th. Passage of the inner to outer. Creative potential. Most stimulating interval. Euphoria and Joy. Harmonically it is at peace. Go to when playing melodically. A clearing and a cleansing. Parallel worlds. Expands in all directions it provokes the widest movement of space. Dane Rugier says in his book, *The Magic of Tone and the Art of Music*, that it has a spiritual function and the power to manifest potential, evoking electricity and the creative mind.	"Twinkle, Twinkle Little Star" "Born Free" "2001 Theme" "Flintstones Theme (descending)" It's in a lot of Brazilian music and a major harmonic in Tibetan and Crystal bowls.
Perfect Fourth (4:3)	This interval is airy, comfortable and evokes feelings of serenity, clarity, openness, and light. It doesn't sound as musical. It's like an ancient trumpet alarm. A 4th is often called suspended; it creates the most tension. It is an unstable interval that likes to resolve to another note. Grand, ta-da, bright, uplifting, joy. Awakening of the heart to control issues causing one to feel unpleasant. Used during medieval times by the church as a means of control to reinforce the heaven and hell concept. Associated with awakening. Paralyzing (on the negative). Stressing toward conclusion. Getting attention. Mental Concepts and Forms. Peace, Peaceful Strength. It is not allowed in some configurations because it would be considered a pagan progression. It was also not allowed by the Roman Catholic Church. Dane Rugier thinks that 4th symbolizes contraction. In Indian traditions it represents a return to the mother.	C to F "Here Comes the Bride" "Hi ho, hi ho, it's off the work we go!" "Hi ho" is a 4th. "Amazing Grace" "Smells Like Teen Spirit" Gregorian chant. Bulgarian voices.

continued

ARCHETYPAL RELATIONSHIPS OF MUSICAL INTERVALS—*continued*		
Interval	**Description**	**Musical Examples**
Major Third (5:4)	This interval is considered to have or possess "great sweetness," manifesting auspicious possibilities. The number 3 is also considered to be a number associated with perfection and divinity. Very sweet interval. It is often the next sweet step. Steady, stability, calm, normal. It is sweet, comfortable, square, and resolved, without a lot of character. Sense of beauty expressed in your external world. "Color tones" in music. Expansive (Yang) Music that brings you into inner emotion and psychic life. Daily life. Majesty nobility Pulls a bit at my heartstrings. Waiting for something else to happen, creating a little bit of stress.	C to E "When the Saints Come Marching in" "From the Halls of Montezuma" "Kumbaya" "Summertime (descending)" Found frequently in church music, country music, gospel, jazz, Baroque music
Major Second	The note is very unstable and wants to be resolved – especially when the notes are going down. Rousing, wakening, hopeful. Somewhat dissonant sounding. This is an interval which creates discord and irritation. However, tends to promote growth and ultimate beauty. Building blocks of life. Patient steps toward larger goals. Connections between things. Complication of life. Patient organization in space and time (Yang).	"Do Re Mi" "Frere Jacques" "The First Noel" Often used in classical music
Major Sixth (5:3)	Sad, sorrowful, weeping. Portal of total opening – feelings of offering yourself to the universe. It is the sweetest interval. Soft and sweet. Dreamy and full of possibility. The flowers open to at dawn. It is the most etheric of intervals. No tension, no wait, no emotional heaviness, very, very pleasing. Has a little more tension to it. Ambitious.	First two notes of "My Bonnie lies over the Ocean" "NBC theme" "It Came Upon a Midnight Clear" "Nobody knows the Trouble I've Seen (descending)" It's in a lot of Gospel Music. 4th's, 5th's, and 6th's are in a lot of Scottish Music (bagpipe music). Lisa Rafel's CD – lullabyes for children. Associated with Lullabyes.

Major Seventh	Most dissonant and uneasy intervals
	Very often is dying to resolve to the next note above. Piercing, sensitive, ready to move. Very sweet when played within a Major 7th chord. It is yelling for resolution. It is called the "leading tone." Provokes growth in consciousness. Has a profound influence physical, musical, subtle, and etheric bodies. It calls for urgency to move into oneness.
	
	Your relationship with the group.
	
	    Major – Through thinking, material things, through works (Yang)
	    Minor – Through feeling, sensitivity, awareness (Yin)
		"Take on Me"
		"Bali-Hi"
		Thelonius Monk played it
		Beethoven played this a lot (first revolutionary jazz musician). Very much the same as the minor 2nd.

ARCHETYPAL RELATIONSHIPS OF MUSICAL INTERVALS—*continued*		
Interval	Description	Musical Examples
Minor 5th or Augmented 4th (sharp 4th interval) – "Tritone"	The "Inverse" or "Complimentary" Frequency The most dissonant because this interval operates outside of the standard Ionian scale pattern or existing tonal circle. This non-conformity creates critical and creative thinking – very energizing and uplifting. Stimulates the left and right sides of the brain. Considered VERY POWERFUL. This note wants to, and often does, resolve to the 5th in music. "It alone, I think, is one of the most powerful intervals that we have," says Kay Gardner. According to standard music theory, this interval is not to be used in music. In fact, it is called Deabolics en Musica (The Devil in Music). Considered to be devils interval . . . summoned demons. Evil in Medieval times. It is not one of the white keys. People were beheaded for playing a tritone. Even in classical music, it wasn't until Wagner and Beethoven came along that they snuck it in. They then called him a romantic composer instead of a classical composer. In gospel and jazz it is used as a passing tone to get to the 4 or 5. Certain scales have a Tritone built in.	This interval is used frequently in horror films and is used in Leonard Bernstein's "Maria" from "The West Side Story." The harmonics in the Tibetan bowls have an augmented 4th. Black Sabbath, Thrash, Hendrix (purple haze), Simpsons, South Park. "Bop" in jazz.
Minor Seventh – Flatted, Dominant, Diminished Seventh	When you play the 7th it has become to mean Flatted 7th. This is a bit dissonant. (There is a tritone in here – 3 whole steps). Deeper emotional music.	"Star Trek Theme" "An American in Paris (descending)" Sacred music, gospel, country, R&B, blues. Bluegrass can also get really deep.
Minor Third (6:5)	Alerts us to seriousness. Needs attention and needs to be resolved. Melancholy (not depression). State of mind that requires some immediate love.	"What the World Needs Now is Love" "Greensleeves" "So Long, Farewell" "Hey Jude (descending)." Slow and serious songs have minor 3rd.
Minor Sixth	One step along in life. Doing one thing after another. Also waiting for something to happen. Ominous. Sacredness. Sadness. Question mark. Confusion. Little bit of darkness or pending doom. Emotional, loving, or the feeling of lost love. State of transformation. Undiscovered. Too close. The opposite is the Major 3rd (on up to the Octave).	C to A♭ "The Love Story" "Here Comes the Bride" "Charlie Brown Theme" "In my Life (guitar intro)" "The Entertainer (leap after intro)" Chopin's music is filled with this interval in parallel motion.
Minor Second	Definitely dissonant. It creates a tense and uneasy feeling that is expectant, anticipatory, busy and mysterious. Within emotions. Psychological; personal relations (Yin).	"Jaws" Theme "Joy to the World (descending)"
Suggested and recommended reading: *The Role of Music in the Twenty-First Century*, by Fabien Maman. *The Temple of Man*, by Schwaller De Lubicz. *Emotion and Meaning in Music*, by Edgar Myer.		

There are a wide range of other scales that have different numbers of notes and different patterns of whole and half-steps. If one note in a scale is altered it completely changes the emotion and vibe. Each scale has its own individual sound, character, and name. The following table shows the step patterns for the most common scales (just to show the wide range of possibilities). When using the interval pattern to count the scale we utilize 1 and 8 to indicate starting and ending points.

Major (Ionian)	W, W, H, W, W, W, H
Natural minor	W, H, W, W, H, W, W
Melodic minor	W, H, W, W, W, W, H
Harmonic minor	W, H, W, W, H, W&H, H
Major pentatonic	W, W, W&H, W
Minor pentatonic	W&H, W, W, W&H, W
Blues	W&H, W, H, H, W&H, W
Mixolydian	W, W, H, W, W, H, W
Dorian	W, H, W, W, W, H, W
Lydian	W, W, W, H, W, W, H
Locrian	H, W, W, H, W, W, W
Phrygian	H, W, W, W, H, W, W

Engineers and producers with an understanding of the scales can help perform tasks such as helping the band tune up and helping vocalists warm up and loosen their vocal chords.

Modes are special scales. Here is a list of modes and their associated feelings:

Perfect vs. Minor Terminology

Certain intervals are very harmonious or consonant and are called *perfect.* These are the first note in the scale, the 4th, the 5th, and the 8th (which is the octave or same note higher). The *major* intervals are usually associated with brightness or happiness compared to the minor intervals, so this why they are called major. Here is the nomenclature commonly used for perfect and major:

- P1
- M2nd
- M3rd
- P4th
- P5th
- M6th
- M7th
- P8th

THE MUSICAL MODES			
Modern Mode name	Root (relative to "C")	Spelling (from c)	Feelings
Ionian	C	C,D,E,F,G,A,B,C	Boosting, Joyful, Sweet
Dorian	D	D,E,F,G,A,B,C,D	Melancholy, Lamenting
Phrygian	E	E,F,G,A,B,C,D,E	Exotic, Invigorating, Arousing
Lydian	F	F,G,A,B,C,D,E,F	Active, Inventive, Fun, Daring
Mixolydian	G	G,A,B,C,D,E,F,G	Heart-opening, Uniting pleasure and sadness
Aeolian	A	A,B,C,D,E,F,G,A	Bittersweet, Sad, Hopeful
Locrian	B	B,C,D,E,F,G,A,B	Continuous, Unresolved, Other worldly

Chords

A *chord* is when two or more notes are played together at the same time. The intervals that make up a chord give you even deeper and more complex feelings. The songwriter normally creates the chord progressions. A producer will sometimes intervene. More commonly a producer might help to set the chord structure of the harmony parts for background vocals.

The main type of chord is a *triad*. This refers to a chord with three notes like C, E, and G which is called a C major Triad incorporating the 1, 3 and 5 of the major scale. When the 3rd is flattened by a half-step it is referred to as a minor triad.

The type of chord structure that is utilized makes a huge difference in how a part is perceived. Different chord structures elicit different feelings. The most common difference is between a major and a minor chord. Major chords give off elation, whereas minor chords give off a solemn or sad mood. A diminished chord is even sadder. A major seventh has an element of anticipation.

Power Chord (sometimes referred to as the economy chord) is only two notes. It is a chord consisting of the Root (P1) and the 5th (P5). It is used in boogie woogie, jazz, gospel, country, bluegrass and great for creating a drone. The unique quality of the power chord is that it provides fullness without being major or minor. Therefore, it can be used in either major or minor musical chords.

All chords are derived from six main chords. The first is called the *tonic*.

Six Main Chords

Major triad	C, E, G	1st, 3rd, 5th (often called perfect 5th)	
Minor	C, E♭, G	1st, minor 3rd (flat), 5th	Sad!
Diminished	C, E♭, Gb	1st, minor 3rd (flat), flat 5th	Sad, sad, sad.
Suspended	C, F, G	1st, 4th, 5th	Wishes to resolve to the 3rd.
Dominant	C, E, G, B♭	1st, 3rd, 5th, flat 7th	Cross between leading and sweet. Bluesy. (Gospel and country.)
Augmented	C, E, G♯	1st, 3rd, sharp 5th	Leading.

The following table shows some other chords that are well-known extensions of those mentioned in the preceding table.

Major seventh	C, E, G, B
Major ninth	C, E, G, B, D
Major eleventh	C, E, G, B, D, F
Major thirteenth	C, E, G, B, D, F, A

You can make any one of these a minor chord by flattening the third interval. You can make them diminished by also flattening the fifth. You can make them dominant by flattening the seventh, and you can make them augmented by sharpening the fifth.

Musicians commonly use *chord charts* and it is really helpful for the producer to be able to work with them. The symbols in the *chord charts* are as follows:

Major – M, Mag, mag, △ – CM, CMag, C△ the triangle also represents major. **1 3 5**

Minor – m, min, – Cm, Cmin, C–, the minus sign is also used. 1 ♭3 5

Augmented – Aug, aug, +, ♯5 C♯5 C+. 1 3 ♯5

Diminished – dim, o, Cdim, C○

Half Diminished – Cm7♭5, Cø

Suspended – sus, sus4. Csus, Csus4

Dominant – b7.,C7

Flat – ♭

Sharp – ♯

Chord Voicing or Inversions

Voicing or *Inversions* refer to moving one of the notes in the chord up or down an octave. So instead of C, E, G, you would have E, G, C. Inversions are used to create smoother chord progression movement where some inversions may feel and sound better than others. As we'll discuss later, they are also very good for creating subtle dynamics.

For example, suppose you have a major triad in C. The highest note in the chord is G, and the lowest note is C.

Visual 64
Virtual Mixer C, E, G

If you invert the chord so the C is played an octave higher, it is called the *first inversion*.

In this case the chord has a little more intensity to it. The top note of the chord always appears louder. If you were to raise the C another octave, it would be even more intense. The higher you play the top note, the more intense it is. It is common to use the first inversion in a harmony part (vocal or strings, for example) to fatten the melody line.

If you were to switch it so the E is on top, it is called the *second inversion* (see Visual 66).

This inversion creates even more tension. Notice how the chord inversions are higher and take up much less space in the mix (see Visual 67).

Closed Chords are chords where all the notes of the chord are within one octave. *Open Chords* are chords that have a further distance between them, often over multiple octaves. In jazz and gospel the 3rd in a 1, 5, 3 chord is in the second octave creating a big sounding voicings.

Visual 66
Virtual Mixer G, C, E

Visual 67
All Chord Inversions

Spreads

The range of the *spread* of the chord creates even another dynamic. For example, a three-note chord could span one octave or three octaves (see Visual 68).

**Visual 68
Virtual Mixer Chord Spreads**

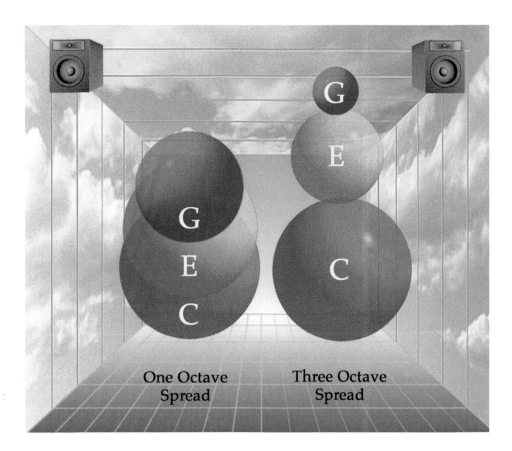

One Octave Spread

Three Octave Spread

The wider spread, the fuller it will sound. It also will take up more space in the mix.

Circle of Fifths

Changing a song to a different key is called *transposition*. The *circle of fifths*, also known as the *cycle of fifths*, serves as a guideline for understanding scales and transposition. Often a producer might suggest changing a key so that it fits the singer's vocal range better. Or, perhaps a different key simply feels better.

Visual 69 shows you the number of flats or sharps for each key. (see next page)

You could simply sit down and memorize the number of flats and sharps, or you can use the chart. Many people derive some spiritual meaning from this circular relationship of all notes and music.

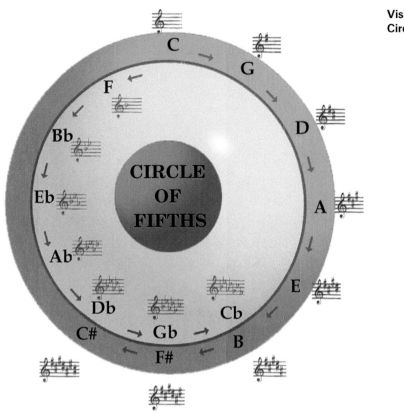

When you change the whole song, it is called *transposing*. When you change a section of the song to a different key, it is called *modulation*. Musicians modulate up to help create an element of drama and emotion in a song. It can also create a very profound emotional dynamic or bring a song to big finish. The most common intervals to modulate up are a half-step, whole step, or fourth step. Various artists, such as Stevie Wonder, continually modulate up a half-step at a time at the end of the song.

Occasionally, songs will also modulate down.

Chord Progressions

Chord progressions are the way the chords change over time. Along with the rhythm chord, progressions create the basic *flow* of the music. The musical intervals referred to are in relation to the root or key of the song (not the previous note).

As Miles Davis so aptly put it, "It's all blues," meaning that American music and even western classical is based on the 1, 4, 5 progression. You can think of all music as a variation of this basic progression. This progression is even associated with the archetypal pattern of the "Journey of the Hero" as described by Joseph Campbell in his book, *The Hero with a Thousand Faces.* The hero starts out at home (1, 4) . . . then goes on a journey (5). Then brings back he or she learned to home (1). This pattern is found through all religions, fairy tales, and myths. It is also the basis of the structure of every Star Wars movie (according to Lucas). There is something very deep about this archetypal pattern within us.

The Motown hit song, "My Girl," recorded by the Temptations, written by William "Smokey" Robinson, but masterfully brought to life by the hit music machine, "The Funk Brothers," is an example of the 1, 4, 5, expansion.

Songwriters and producers like Pharrell Williams, Cee Lo Green, and Bruno Mars (Mark Ronson) all use the 1, 4, 5, and the expansions in very similar ways. Expansions are used to extend the tension and release of the 1, 4, 5 by using additional degrees of the major scale to lengthen the progression. For example, '50s doo-whop and rock-and-roll used the 1, 6, 2, 5.

The use of substitutions outside of the scale may also play an important role in creating tension, movement, color, and emotion. The important thing to remember is that the 4 and 5 are the degrees in the scale that take you back home to the 1, thereby creating a sense of completion. This concept is found in almost all music. Some music will actually play around the edges of this concept. For example, jazz commonly avoids the home note (1) in order to create tension. Whereas, Sound Healing music commonly goes back home. Sometimes an entire song stays on the home note – this is referred to as a drone.

Volume

In sheet music, symbols are used to show how loudly things are played. These are the volume dynamics played by the musicians, as opposed to the volume settings of a fader on a mixing board. Although you can simulate musician dynamics by using a volume fader, you won't get the emotional dynamics that normally accompany the rise in volume. It is important for the producer to pay close attention to the volume dynamics in order to get them the way you, as the producer, would like.

Really great musicians have mastered the use of volume as a major part of their performance. However, when a whole band with magical chemistry work together using volume dynamics in synch, the effect is dramatic and can create dynamics that are life changing.

What follows are typical musical terms for the volume at which an instrument is played.

ppp	Pianississimo	Very, very soft
pp	Pianissimo	Very soft
p	Piano	Soft
mp	Mezzo piano	Moderately soft
mf	Mezzo forte	Moderately loud
f	Forte	Loud
ff	Fortissimo	Very loud
fff	Fortississimo	Very, very, loud

Accents

sf	Sforzando	A sudden strong accent
sfz	Sforzando	To force
fz	Fortezando	To play loudly immediately
fp	Fortepiano	To play loudly, then immediately softly
>	Decrescendo	To slowly get softer
<	Crescendo	To slowly get louder

Using the Virtual Mixer concept, musician dynamics are normally shown as the brightness of the sphere. (Volume faders are shown as front to back.)

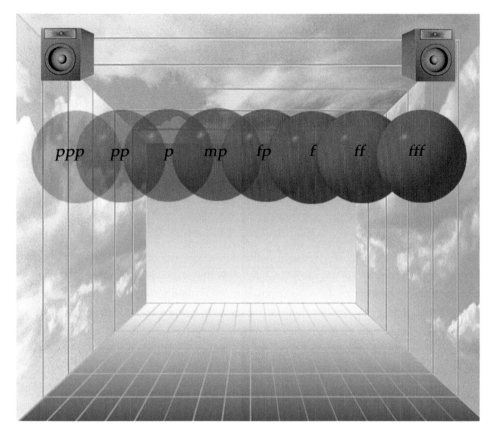

Visual 70
Musician Dynamics in the Virtual Mixer

Song Structures and Musical Charts

There is a wide range of ways to display music, other than using normal sheet music with staves. You will often come across these in the studio. You should be aware of some common terminology.

Phrases

A *phrase* is a group of notes or lyrics that form a cohesive musical idea. Phrases are smaller sections within a section of the song (such as a verse or chorus). They are often used by everybody in the studio as a reference to discuss a part or to tell people where to start playing or singing. They can also be used to simply let people know where to breathe.

Repeat Signs

Repeat signs are like loops in a sequencer. Instead of writing the same section over and over, symbols are used to indicate what sections to repeat in the song. When two dots are placed before a double bar, this simply means the musician should go back to the beginning of the song and repeat the music again.

In a sequencer, repeat signs are commonly displayed as you see here.

Visual 71
Repeat Signs on a Staff

If you want the loop to start somewhere other than the beginning of the song, then place a

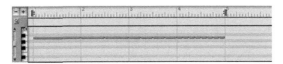

Visual 72
Repeat Signs in a Sequencer

double bar with the two dots to the right at your starting point. In sheet music you simply write "Repeat *x* times" to indicate how many times to loop the section.

Visual 73
Loop on a Staff

There are symbols that allow you to skip the last measure(s) of the loop and continue on with the song. These are called *first* and *second endings*.

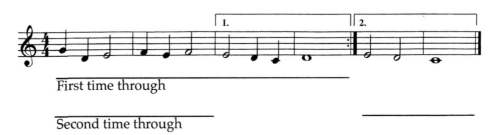

Visual 74
First and Second Endings on a Staff

You can see that musical section will repeat at the end measures, but the second time around the fourth measure is skipped and the rest of the song is played.

Shorthand Chord Charts

Sometimes the names of the chords are placed above the staff with the melodic line written. This concept is the basis of *Fake Books* and *Real Books* that don't provide the entire arrangement of all of the instrumentations but do provide the entire song structure where the vocal or instrumental line is written on the staff with suggested and/or alternate chords. The symbols used were outlined previously.

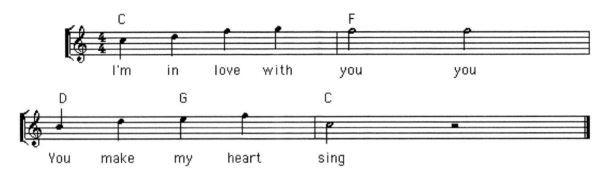

Visual 75
Chords Above the Staff

Sometimes the chords are simply by themselves with the lyrics as below.

```
C                              F
I'm     in     love   with     you          you

D               G              C
You     make    my     heart   sing
```

Visual 76
Chords with Lyrics

Shorthand charts can be very useful during pre-production for mapping out a song. They are commonly used in recording sessions so that everyone knows what to play without being overwhelmed by the detail of written music on the staff!

When mapping out a mix, charts allow the artist, producer, and engineers to make notes where certain kinds of emphasis and effects will take place in a song.

PART 5

The Production Process

The first half of this section is on pre-production, and the second half is on producing while you are in the control room and studio.

Chapter 5: The Pre-Production Process

As previously mentioned, to obtain the job of producer you must first produce a band. Therefore, if you are trying to break into the field, you must find a band that will let you produce them. We'll discuss specific techniques for finding a band later, but for now it is important that the band meets the following criteria:

- Find a band that you like.
- Find a band that is good. It is of little use if you produce a band, and even do an incredible job, but then you don't want to play it for anyone because the band sucks.
- Find a band that you can connect with – in other words, a band that will allow you to produce them. Not only do you want to find a band that you like, but you want them to like you. It's about respect. In the beginning, it can be difficult to command the respect you need because of your limited experience. The trick is to show humble confidence. Don't be overly cocky; everyone knows you are learning. But do show confidence that you know what you are trying to obtain, and you know how to get it. The key is to be able to clearly explain exactly how you can help the band from the beginning. And then, keep that in the forefront of your mind the whole time. Later, we'll provide many tips and techniques for how to work effectively with a band.
- Find a band with at least a little bit of money. They have to be able to pay for the studio time, but it is also good if they have a little extra money to do some of the things we'll be discussing later.

We'll discuss more tips for finding a band near the end of the book, once you have picked up more techniques and gained the confidence to go out and do it yourself.

Remember, for your first band, it is not critical that you get paid. Think of it as a learning experience. They might also recommend you to others, and they might use you for further projects.

Setting Up the Meeting

Once you've made an agreement to produce a band, you schedule a pre-production meeting. Try to set up a meeting at a time when all of the members can make it. Again, it is important that every single person in the band is present. As a last resort, you can set up the pre-production on nights when the band is already scheduled to rehearse, although it is a shame to take over the band's rehearsal night. (They might need it!)

Starting with your initial contact with the band, you want to begin establishing a trusting relationship. The big problem you are up against is a fear that is rampant in our society – fear of producer. All musicians have it. If you are a songwriter, I bet you know what I am talking about: You don't want a producer telling you what to do. I know that I would think: "It's my music. I know what sounds good. I don't want someone else telling me what is good. I don't even want to have to waste my time arguing with someone about what is good. It's my music. I *know* what sounds good!" (You can imagine that people can get pretty touchy.) Almost everyone experiences producer fear to some degree.

As the producer, you must work hard to help the band overcome this fear of producer. The first step is to contact the bandleader and get each member's phone number. Then, call each one up and introduce yourself. If you haven't talked to each person in the band, then they are quite likely going to be defensive. You can imagine the leader of the band going to the rest of the players and telling them, "We're going to have this meeting, and a producer is going to be there." Can't you just hear the rest of the band members saying, "How many hits does he have? He's going to produce us and he hasn't had a bunch of hits!" Therefore, it is critical that you connect with each person in the band to begin to put his or her fears to rest.

When you talk to each band member on the phone you want to let him or her know that you are not going to tell them what to do, and that if you come up with any ideas they don't like, they don't have to use them. Tell them your main goal is to simply collect all of their ideas and not forget them, so they can be used in the project. Let them know that you will be writing everything down to keep track of all the ideas. Ultimately, you will evaluate these ideas and talk with the band about how they all fit together to create an overall piece of art.

It is important to explain to the band members that your goal is only to bring out the best in what they already do – individually and as a band. Explain to them that your primary concern is coming up with what *they* think is the coolest thing, because that's what is going to make the project the most cohesive.

Even famous producers rarely tell people what to do. If you get on the wrong side of the band, it's over; you've lost them. The basic idea is to create a team effort so everybody works together. If they are upset with you, they aren't going to play well. (Unless they doing angry songs – for a punk band, it could be cool. Get them pissed off, and they will play better!) Initially, though, you want to try to get rid of the "fear of producer."

During your initial phone conversations, you also want to emphasize that the pre-production meetings are serious – it is critical that they be there for the meetings from beginning to end. That way people won't say, "Oh, I've got to meet my partner. It's okay if I leave an hour early, huh?" Let them know that it's very difficult to make critical decisions if every single person in the band is not present.

You should also tell the band members that if they are going to be successful, not only do they have to be there (and on time), but they also have to be sober.

In summary, set up meetings with everyone for pre-production, get everyone's phone numbers from the band leader, and talk to everyone in the band to put them at ease.

Homework Before the Meeting

Before you meet with the band for the pre-production meeting, you need to do some homework. Keep in mind the following guidelines:

- **Keep a notebook for each project.** The producer should always keep a notebook for each song in the project. A three-ring binder works really well because you can insert copies of the lyrics, song maps, and any other notes that you are given or ideas that you might have written on toilet paper or other pieces of scratch paper.
- **Get the songs recorded.** First, you need to get all of the songs that the band or group might want to record. Often, they will already have a demo of the songs. If not, go to their rehearsal, and record all the song on your phone (quality is not critical here).
- **Get all the lyrics on paper.** Hopefully, you can simply go to the songwriter, and he or she will have the lyrics already written down. If not, hopefully you can transcribe the lyrics from the recording. If not, get the songwriter to help you. Get all the lyrics on paper, make copies, and bring those copies to the pre-production meeting.
- **Make song maps of the song structures.** Map it all out so you know where each sound is playing throughout the song. You can't produce somebody's music if you don't know the song structure and instrument layout in detail. You aren't going to have any credibility if you don't know the band's song. These guys have often been playing the songs for ages. They might have been playing them live in concerts and clubs, and they know the songs inside out. You need to get to know them, too. You need to know the structure of the song so you can discuss what to do in each section.

Take the recordings of their songs home, map out the order of the sections of the songs, and list all of the tracks. Again, a spreadsheet program such as Microsoft Excel works quite well. Map out the order of the sections – intro, verse, chorus, lead break, bridge, vamp, and so on. It would be nice to have actual times for each section, but it isn't necessary at this point. Map out all the instruments – kick, snare, hi-hat, toms, bass, guitar, and so on – and where they are playing throughout the song. Draw a line whenever they are playing. The chart should look something like Chart 4 (see next page).

Note that in our example, the song begins with a sound effect. The first verse is actually quite sparse with only drums, bass, key 1, guitar 1, and vocals. The percussion drops out in the first verse. The first chorus adds another vocal, background vocals, guitar 2, key 2, strings, and a sound effect. The first tag then drops all of this out except for the strings. Horns, key 2, percussion 3, and sound effects are added underneath the lead guitar part. The same density is repeated in the rest of the verses and choruses. The lead break again breaks down most of the instruments but a different percussion part and horns are added. The bridge has almost the same density as the choruses minus the main guitar and key. The breakdown adds a whole other keyboard.

SONG MAP

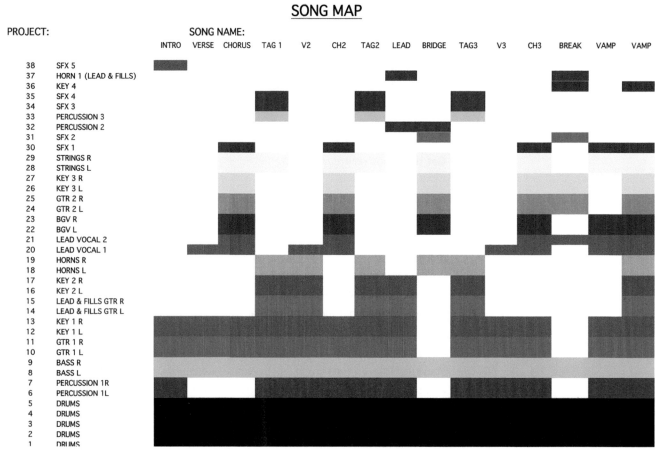

Chart 4 Song Map

As you can see from this chart, we are going to need 44 tracks to record this song. Most digital audio workstations don't have that much power, so we will either have to lose some tracks or plan to mix some tracks together at some point in order to free up some tracks.

- **Go through each song and critique each of the 13 aspects.** Take the recordings home and critique each of the 13 aspects of a recorded piece of music for each song. Chapter 7, "Structuring and Critiquing the 13 Aspects," provides the details of each aspect and how to critique it. (You also will find a chart of all the homework to be done before the meeting in Appendix A, "Homework to Be Done before the Pre-Production Meeting.")
- **Schedule the order of the recording of the songs and the musical parts.** Sometimes it is good to start with an easy song. An up-tempo song can help get everyone's energy going. Very slow songs tend to be much harder to record and take more time, so save them for when everybody is warmed up. Normally, you should try to record the basics (drums, bass, and other basic rhythm instruments) first. You don't want to set up and mic a whole drum kit more than once. Also, because the order of the songs often depends on the specific instrumentation, you want to record all of the songs that have similar instrumentation in them, one after another. You don't want to have to re-mic instruments.

The order in which each instrument is recorded is very important, whether you are dealing with chamber music for small orchestras or contemporary music, such as neo-soul (R&B), jazz, country, or pop. Having this information available beforehand also helps the engineer set up in advance.

First, decide which instruments you will record during the basic tracks (the first recording of parts together). It is often good to get as many of the players together as possible to get the feel of the band, even if you are planning to overdub (re-record) each part separately later. Often people play much better when all the parts are there because they play off of each other. Of course, the ability to do this is limited by the number of isolation rooms in the studio. In jazz (and some other styles), it is often good to isolate a singer so you might be able to keep his or her live performance. Sometimes recording the original vocal with the whole band is the best, but normally vocals, lead instruments, and horn sections are recorded as overdubs.

Figure out exactly how many instruments you will be recording during the basics. Determine precisely where each instrument will be placed in the recording studio, and then figure out the order of the overdub recording. Sometimes it is nice to do the background vocals to scratch (rough) vocals, so the singer doesn't feel lonely when the real lead vocal is recorded. However, if the background vocals are going to precisely mimic the lead vocals, then the lead should be done first.

Occasionally you can do them all at the same time, but it makes it more difficult to punch-in and fix individual parts. It is good to have complete isolation so you can punch-in individual lines.

As a rule of thumb, always begin with the rhythm section (bass, drums, guitar, and keys) if possible. If for some reason you cannot get these four pieces together at one time, start with the drums and bass. It is extremely important that these two instruments are locked together, laying down the foundation for the guitar and piano – particularly if you are recording dance music or ballads.

People Skills

Pre-Production Meeting Introductions

Once you get into the pre-production meeting, start by once again putting the band at ease around their fear of producer. Reiterate what you discussed on the phone by saying things like, "I'm not here to tell you guys what to do. I only want to help bring out your own best ideas. If you don't like my ideas, you don't have to use them. It won't hurt my feelings. What I want here is an environment of creativity. I want a proliferation of ideas to flourish. I want everyone to feel free to come up with and present ideas because this is a creative meeting." The point is to simply let them know that you are not going to tell them what to do; you are sincerely interested in their ideas. Besides, it's their music.

Deciding on Deciding

Because pre-production meetings are fraught with so many possibilities for conflict, it is important to be very careful and take precautions to keep everyone from getting into arguments with each other.

First, you might lay down the decision process. How you do this depends on the egos in the band. Normally, the best way to do it is democratically – the most votes wins. But sometimes, if you have big egos or extra-sensitive people, you might want to set it up so that the vote has to be unanimous in order to make a major decision. In this case, everyone has to agree before you do anything. However, the truth is, it is really hard to work this way. It's much harder to get things done. Ideally, it is best to set up the decision process based on the democratic system.

The problem arises when you have one person who is the leader of the band, and it has already been established that he or she always makes the final decision. It is their band. Feel out the band from the beginning and try to see what the hierarchy or power structure is. You don't want to try to make it democratic if it is not already democratic because you can get hurt. You're not there to be a therapist. Once you find out what power dynamics are going on in the band, adjust your approach to fit. If one person in the band is controlling everything, make them your best friend. Ask their opinion on everything. Then, smoothly, start asking everyone else's opinions to try to balance things out.

It is important to be aware of this pecking order because it can become a serious issue later, when you are negotiating with a record company. Record companies would often prefer for one person to own the band and the rest to be employees. It is much easier for them to deal with one person than an entire group of individuals when it comes to negotiations. Therefore, you might help the band to begin making this decision in order to minimize any blowups later.

Passion and Referees

In these production meetings, quite often you will find that people can get very passionate about their ideas. We used to be annoyed by this intensity, but we have learned that the more passionate people are about their ideas, the better the project normally turns out. However, you can have fireworks if you have two people who are both passionate about their ideas and they don't agree. And that's when bands can easily break up.

There is a very large potential for a band to break up during these meetings. As you go through each component of the music and performance with a fine-toothed comb, it often brings up many opportunities for bands to air any animosities. It's like therapy: You are diving into places where people often have deep, pent-up anger. The truth is, there is often a lot at stake. The success of the project could mean the difference between the band making it or going back to jobs they hate – or in the worst case, living on the streets. Therefore, people can get very intense with their ideas. Again, though, we have come to appreciate this passion.

There was an interview with Neil Schon of Journey in which he said that every time the group was in the studio, they fought like cats and dogs. But because everyone was so passionate about their ideas, all of the projects came out incredible. He said that he then got a record deal to do a solo album, and when he went into the studio, everyone asked him what *he* wanted. He said it was a lot easier and a mellower session, but the final project didn't come out nearly as incredible as when everyone was screaming at each other.

We once co-produced a new age project of totally beautiful music. Everybody in the room had quite divergent ideas and they were seriously passionate about them. Things often got quite heated in that control room. It was the mellowest music in the world, but the most stressful session possible (or so it seemed). However, the project came out to be quite unbelievable.

What happens is, the project takes on that intensity. You get the best of the best ideas, and you normally end up with much more depth because other ideas might still get incorporated and placed in the background in the mix.

We realized years ago that as recording engineers, if you have someone who is unusually intense, anal, or a perfectionist, the project always ends up coming out better. Therefore, we're not only *not* annoyed by these people anymore; we often look forward to working with them!

The problems start when the band members go overboard and start fighting. People often have really strong feelings. Again, it's their livelihood; it's their whole career. It's their art, and it's their heart. So what you want to do is say up front, "You know, guys, sometimes it's just not worth it." People are going to have really strong ideas, and the truth is, sometimes it's just not worth it because you don't want the band to break up. So let them know; tell them, "Sometimes you guys are going to have to compromise."

One of a producer's most important jobs is to function as referee. You don't want the band to break up, and you don't want them to lose their camaraderie and cohesiveness. Besides, when it comes down to it, this is what you are recording – the vibe between the band members! Do whatever you can to resolve conflicts and help people compromise. The key is to keep reminding them of the higher goal – working together to create great music and become successful, or simply light people up.

Producer Passion

Not only do band members get testy and passionate about their ideas, but it is easy for you, as the producer, to become overly passionate and overzealous. Sometimes it's not worth it for you to go overboard. And it is easy to sometimes go overboard. As a producer, you often have even more at stake than the band. As a recording engineer, it's easy to say, "Well, it's not my fault that the band sucks. I'm getting paid anyway." But as a producer, you start taking it personally because your name is going to be on the album. You can't put a disclaimer on the album (although we have heard of people asking to have their names taken off of albums).

It's very easy to get really touchy. When you start thinking about how important this project is to your own career, it is easy to become the controlling producer. You might have to remind yourself, "It's not worth it." If you lose the confidence and respect of the band, you can no longer be effective. If you upset the band, they very well might stop listening to you and your ideas altogether. Sometimes you just have to compromise.

> **It is important to have strong opinions,**
> **but don't get too attached to them.**
> **Sometimes you simply have to let go.**

How Important Is It?

When you are critiquing any component of a song, make a judgment call about how important it is on a scale of one to ten. If it is above a five, then it is important enough to push for with some vigor. No matter how strongly you feel, always remember that sometimes it is not worth it. When the importance is less than five and your feelings are not very strong, it is much easier to have no attachment to the outcome. The following table provides a list of levels of importance.

Eight and above	Put together your case for why it should be this way. Give it your everything, because at this level you are convinced that it is correct. (Do listen to other ideas and see if they change your mind, though.) Still be willing to let go if you don't convince everyone.
Five to seven	Make your case and give it a good shot, but be willing to give it up in the end if it doesn't fly.
Three to five	At this level, you might have an inkling that your idea makes sense, but you don't have a complete case that you can put together to present to people. Put feelers out there to see whether your ideas make sense. Present your ideas to the group and open it up for discussion with the goal of trying to figure out whether the idea makes sense.
One to two	Present your idea off the cuff and see whether there are any takers or thoughts that might develop the idea into something stronger.

Consideration

How many producers does it take to screw in a light bulb?

I don't know . . . what do you think?

**It's all about creating an atmosphere
where a proliferation of creative ideas can flow –
a place where everyone feels safe
coming up with and presenting ideas
without fear of recrimination.**

It is critical to keep a proliferation of creative ideas flowing. It is critical to give complete respect and consideration to anyone's ideas regardless of your initial impressions. Whenever you are making comments about anyone's music or contribution to a song, wear kid gloves. It also is helpful to end such a statement with, "I don't know; what do you think?" This is not only considerate; it opens the door for the musician to give his or her opinion. It can also help desensitize any overly sensitive egos.

When critiquing a performance, you might say something like, "That was really good, but I think you could do it better." Or, "That was great. Have you ever thought about throwing in . . .?" It's always good to sandwich criticism with sweetness.

The truth is that some very famous producers (and even some who aren't so famous) will not spare the feelings of anyone in the session. Their goal is to get the best quality at any cost. However, costs may include loss of a creative atmosphere or a bad performance because a musician is steaming. Respect and consideration help keep the energy positive and flowing in a session.

Often, the better you know someone, the more you can criticize him or her. With bands we've worked with over the years, we've got to a point where we can say things like, "How the hell could you possibly think to play something so stupid? I don't know, what do you think?" without any bad feelings at all. When you really know someone, the respect is understood.

Time Constraints

It is important not to turn the pre-production meeting into a complete workshop. You could easily end up spending a year working on all the possibilities available to each song. As the producer, it is critical for you to define only one or two (or so) directions in which the project should go. Then, head in those directions – otherwise, you may never finish.

The normal process is to collect, collect, and collect ideas. You might even have the band or a musician play the song or a part in multiple fashions, possibly even with a variety of different instruments or instrument sounds. This is a good procedure to simply get the creative juices flowing, and also to develop as much content as possible. This is the easy part, because most musicians will play until their arms fall off. The big problem comes when you have to really make some serious choices about what the final arrangement and mix of sounds will be. It is all too easy to try an infinite number of combinations. One of the biggest faults of a producer is not knowing how to budget time to actually get the project done.

Chapter 6: Dynamic Flows

The first task after getting the song mapped out is to find the dynamic curve of the song. Dynamic curves exist in almost all music, it is responsible for the emotional, intellectual, and spiritual activation of a song. Some refer to this concept as peaks and valleys in musical story telling. There are some basic archetypal "flows" that songs and genres follow. Here are some examples:

The Dip

Meet people where they are, bringing them down into the depths, and then bringing them back up a bit at the end of the song. This example would be a song telling a story very close to simple life that has dips and almost levels out. Genres and types: Country, '70s Pop and R&B. This curve is a very common curve used in the Sound Healing field – meet people where they are at, bring them into peace, then bring them back up so they can function and drive.

Mellow Out

Bring them down, down, down to a place of deep peace and presence. This example moves us to a constant space of connection to bigger pictures. Genres and types: spirituals, classical, jazz ballads, Ancient Yoruba, Asian, Native American, Pacific Island, inspirational, toning, higher-consciousness music. In Sound Healing it is especially helpful for those with anxiety.

The Stairway to Heaven

More and more activating. Genres and types: gospel, inspirational, ragas, some hip-hop, salsa, Afro-Cuban, the music of Pink Floyd, Earth, Wind & Fire. In Sound Healing this curve is especially helpful for depression.

The Mountain Journey

More and more activation, then mellow out. Genres and types: folk, love songs, hymns, some hip-hop, mariachi, and bluegrass. This curve is helpful for those who are overwhelmed.

Stability

Very few dynamics. Perhaps while some parts are becoming more activating, other parts are getting calmer . . . always balancing each other out. Genres and types: electronica, trance, house, dub and crystal bowls. This can be good for anyone who is on an emotional roller coaster, or who is very sensitive.

There are many aspects of sound and music that create the ups and downs for each of these curves.

Sound and Music Dynamics

	Calming	*Activating*
1.	Even Harmonics	Odd Harmonics
2.	Complex (Rich) Harmonic Structure	Pure Sounds
3.	Harmonic Intervals or Chords	Dissonant Intervals or Chords
4.	Soft Volumes	Loud Volumes
5.	Fewer Instruments	More Instruments
6.	Low Frequencies	High Frequencies
7.	No Tempo; Slower Tempo	Faster Tempos
8.	More Sustain or Legato Notes	Staccato or Short Duration Notes
9.	Simple Melodies	Complex Melodies

10.	No Harmony Parts; Unison Harmonies	2- or 3-Part Harmonies
11.	Simple Song Structure	Complex Song Structure
12.	Calming Nature Sounds	Activating Nature Sounds
13.	Sparse Mix; Reverb	Full Mix; Trippy Effects
14.	Vowel Sounds	Lyrics
15.	Male Vocals	Female Vocals
16.	Repetition	Change

Let us explain in more detail.

1 Even harmonics are instruments or sounds that are warm and calming. The classic example is that of the harp. Any nylon string instrument, like the classical guitar, will also be more mellow. Instruments made of wood tend to be warmer also. The tongue or log drum is a great example of an even harmonic instrument. Odd harmonics are instruments or sounds that are edgy or activating. These include bagpipes, gongs, sitar, and any reed instrument (sax, clarinet, oboe). Also things made of metal create more odd harmonics. Tibetan bowls are mostly odd harmonics. Finally, things that are played really loudly are full of odd harmonics, for example a scream, hitting a drum really hard, or distorted guitar set to "11" on the volume knob.

2 Sounds with very few harmonics are called *pure.* This would include a sin wave tone generator, tuning forks, crystal bowls, flutes, and some bells. A sound with a large number of harmonics is called *complex* or *rich.* This would include the voice, piano, violin, cello, or any large instrument. Generally, pure sounds tend to be more activating while rich sounds tend to be more mellow.

3 Beautiful musical intervals and chords are called *harmonic* or *consonant.* Annoying intervals or chords are called *dissonant.* Obviously the harmonic chords are more calming.

4 Soft volumes are more calming than loud volumes.

5 Fewer instruments are normally more calming, but if one instrument is extremely edgy and activating, then more instruments will actually mellow it out by making the one instrument less prominent.

6 Based on the research of Alfred Tomtis, low frequencies calm the body and high frequencies activate the mind.

7 About the only thing that doesn't create a tempo is tone. Even crystal bowls often have slow oscillations that actually do create a tempo. Obviously slow tempos are more calming that fast tempos.

8 Long sustain notes are notes that last a long time tend to create more peace in us. Unless a sound is annoying, "Peace is a frequency humming consistently." Generally, if you have a lot of sustain notes the tempo is slower anyway. Short duration staccato notes activate both our body and brain. Especially when you have many, one after another.

9 If not boring, simple melodies are much more calming than a busy melody.

10 Harmony parts simply mean singing or playing an instrument more than once at different notes simultaneously. A unison harmony means that more than one person sings the exact same note. This requires multiple people in a live session, but can be done with multitrack recording in the recording studio. Two- or more part harmonies are generally more

activating. It is important to point out that just because something is activating does not mean that it can't still put you in the zone. Many harmony parts are incredibly beautiful and can completely activate you into a state of calm bliss.

11 The song structure is the order of the sections of the song: verse, chorus, bridge, lead break, and so on. A crystal bowl playing by itself is the simplest song structure you can have – it has only one section for the whole song. A song that has many different sections tends to be a little more activating.

12 Nature sounds touch a primordial deep aspect of our being. It's the way God built it. Nature sounds run the full gamut from calming to activating. Mellow crickets, bird chirps, gentle rain, water lapping on the shore, or a small bubbling brook are quite calming. However, an edgy "caw" from a crow, torrential rain with lightning and thunder, huge crashing ocean waves, or roaring river or waterfall are all quite activating.

13 A sparse mix is a song that has very few instruments. A full mix is packed with sounds. It is a very powerful dynamic to build from one instrument to many over time, then do the opposite until you again only have one instrument.

It is also very effective to actually create a higher chord inversion every four bars or so in order to move the music slowly higher and higher.

Or, you can also have the chords cover more and more octaves every four bars or so, to create more and more fullness . . . and vice versa to create more calm.

A sparse mix can also mean an audio mix that has no effects added in the recording studio while mixing. Reverb and long delays tend to also mellow things out. Whereas a deep flange, distortion, or preverb (backwards reverb) can be extremely activating.

14 When you have lyrics in a language that people can understand, another side of the brain kicks in. This side of the brain understands words and is also the judgmental side. Therefore, when you have songs that just use vowels or are in languages that we don't understand, it is actually mellower.

15 Male vocals are only more calming because they are normally lower in pitch and have more harmonics. However, of course, there are many very activating male vocals, and very mellow sound female vocals.

16 One of the most important aspects is repetition versus changing melodies and rhythms. When something repeats over and over (like trance, chant, or even mantra) the left brain goes to sleep because it gets bored. No change, I'm going to sleep. This allows the right brain to go very deep or get really high. On the other hand, if the song has many changes throughout the brain stays awake because it doesn't know what is coming next. This also creates new neural pathways in the brain and is good for learning (it's called the Mozart Effect). Channeled or Ad Lib music is very activating at first, because you never know what is going to happen next. Therefore, it demands 100 percent presence. However, once you are present with the channeling they can take you deeper and calmer than ever.

For each of these, the most calming effect is to slowly change from an activating parameter to a calming parameter. For example, smoothly slow down the tempo until it is really slow. Same for activation . . . if you really want to get people amped up slowly change from a calming parameter to a more activating parameter.

However, the most powerful of all is to use all parameters at once. If you *really* want to get someone calm, use everything in the left column. If you really want to get someone activated, use everything in the right column.

It is really important to map out the dynamics curve of the song in the beginning. Then you now have the tools to make the song even more dynamic – if that is what is appropriate for the song and its message and intention. With these dynamic tools you can take people to the sky . . . or to the peaceful depths of their soul.

Chapter 7: Structuring and Critiquing the 13 Aspects

Once you've got the decision process in place and you have put the band at ease, you can begin to critique and refine each of the 13 aspects of a recorded piece of music with the band. You need to let the band know what you are about to do so they aren't in the dark. This process can easily take three hours per song, and it is normally best to work on one song at a time.

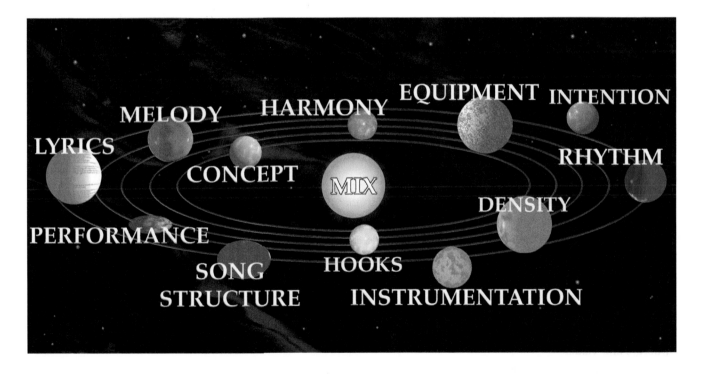

Visual 77
The Solar System of the 13 Aspects

Pull out copies of a page that outlines the 13 aspects of a recorded piece of music. Tell the band that you're going to work together to critique each one of the 13 aspects.

As an engineer working with a band in the recording studio, there is not much you can do to fix and refine many of the 13 aspects (other than critiquing performance). But in pre-production, you refine things before the recording session, before you are paying for every hour of studio time.

Each of the 13 aspects is organized into subsections in this chapter as follows:

- **Structure and Values**. This section provides the basic definition and information on the details of an aspect. It is helpful to understand this information about an aspect so you can critique it. These are common values that producers utilize in refining the aspect.
- **Homework**. This section covers details you need to critique a particular aspect before going into the pre-production meeting. Take copious notes. When you do your homework, make your list of items to bring up in the meeting for each aspect.
- **Approach in the Meeting**. This section details how to approach the band during the pre-production meeting to critique and refine each aspect.

Here's the full list with short descriptions.

- Concept
- Intention
- Melody
- Rhythm
- Harmony
- Lyrics
- Density of Arrangement
- Instrumentation
- Song Structure
- Performance
- Mix
- Quality of Recording and Equipment
- Hooks

Aspect 1: Concept

The *concept* or *theme* can be defined as a combination of the other 12 components. It is not the style of music; it is the overall message that the band is trying to communicate. It is the essence of the song, or the primary message. It is also known as "the overall vibe," "the mood," "the flow," or "the aura," depending on your perspective. It is often defined by the lyrics. If there are no lyrics then it is based on the type of emotion that the song invokes. It is usually defined as the feeling or idea that is conveyed most consistently and strongest in the greatest number of the other aspects. Once you have the concept down, the rest of the parts often fall into place.

Structure and Values

There are two aspects of "concept" to critique. First, is the quality of the concept. Second is the cohesiveness of the concept.

The quality of the concept is a subjective determination that can be judged in many different ways. Some people judge the quality based on whether it is memorable. Often, really stupid concepts make millions of dollars – not because they have any meaning but because they are wacky. Al Yankovich is a good example. There are many songs in which the concept has no value but it is simply fun or interesting. There are many songs where the concept is about partying or just having a good time that have been major hits. Then there are the songs where the concept has some deeper meaning. Perhaps it helps to put a perspective on some situation that is going on in your life, or simply touches you in a really deep way.

The deepest concepts will actually touch the largest numbers of people. Generally, any concept about love has the potential to touch masses of people. There have been a huge number of songs about lost relationships. We often look for songs that have the potential to really affect people and possibly make positive changes in their lives.

The second aspect of "concept" is about cohesiveness. Songs vary in the consistency or cohesiveness of the concept. In some songs, the concept is quite strong and cohesive, while in other songs it might be nonexistent (the concept could be "no concept"). Most people focus on the lyrics when discussing the concept, but when there is a concept, every component should support it. At the very least, each component should not detract from the central idea. For example, if a guitar part doesn't seem to fit the feel that the rest of the band has already established, it should be changed. If any one of the other 12 aspects has something in it that doesn't fit, refine it.

Discovering how each aspect relates and contributes to the cohesiveness of the whole song can often reveal inconsistencies and deficiencies that might need to be fine-tuned. Often a part might need to be taken out or played with a more appropriate feel. However, you should be careful to detect creative ideas of genius that could easily be misconstrued as inappropriate. Even more important, such a detailed analysis can sometimes provide inspiration and lead to the creation of new ideas.

A good example of adding a part to make the concept stronger was done in Pink Floyd's "Teacher, Teacher" ("We don't need no education"). Somebody had to actually think, "Oh, let's get 100 school kids to sing this." Basically, this idea was a result of the concept of the song. Having 100 kids sing the line made the concept stronger and more unique. In this case, they added a new harmony part and new instrumentation (the children).

As an example of utilizing each aspect to make the concept more pervasive, consider the example of a song about love relationships. Here are some examples of what you might do for each aspect to boost the concept. Of course, these ideas are intended only to show you possibilities – each song is different, and these ideas won't necessarily fit any other song. Come up with your own ideas that fit a particular song.

- **Melody.** Make the melody sweeter by making it simpler, allowing more space in the music.
- **Harmony.** Use female singers singing chords that are very sweet. Add subtle oohs and ahhs at various places.
- **Rhythm.** Make the rhythm slower, perhaps. Make it simpler and less busy.
- **Lyrics.** Critique each line, looking for clichés. Look for the lines that really touch you and perhaps have them repeated.
- **Density.** Consider breaking down the arrangement to only one instrument at the sweetest moment in the lyrics and song. Add lush orchestration if the emotions get very deep and intense.
- **Song structure.** You might simplify, or add a bridge section that really goes into deeper feelings.

- **Instrumentation.** Add a section using a flamenco or nylon string acoustic guitar. Add lush strings.
- **Performance.** Think of ways to perform the lyrics or a guitar solo that are heartrending. Have the singer think about his or her mate (or someone they loved at one time). Have the singer sing it "airy," adding a lot of air sound to his or her voice.
- **Equipment/quality of recording.** Use good-quality equipment.
- **Mix.** Mix with lots of long delays and reverb. Create a cavernous space for the bridge section that goes deep into the heart!

Because you know the concept of the song, you can use real-life experiences to add to the song. Suppose you go to a party and someone says something about recently breaking up with his girlfriend. He comes up with a perspective on the whole situation that you realize you can actually use in the song because it totally fits.

Or, you might have keyboard and guitar parts "fighting" each other musically. Pan one full left and the other full right. Then, as they finally get together and the guitar and keyboard begin playing the same parts, you pan both of them to the center – totally together in love and harmony. They're in love. They worked it out. Now the music fits the concept.

You can even go beyond the 13 aspects and think of events and marketing concepts that will create more of a market for the song. Sting did it by getting Jaguar to use his song "Desert Rose" in a commercial. The commercial broke the song! Occasionally, songs seem to fall in line with a particular product. You might consider contacting a manufacturer or business to see whether they might want to use the song for their advertising.

As another example, suppose you are producing a song about anti-violence. Go to a local anti-violence group and see if you can get it on the radio as a public service announcement.

You also can think of ways to tie the concept of the song into events. For the anti-violence example, you could hold an anti-violence concert. You might even do a whole grass roots campaign on anti-violence. Get radio stations to do interviews. You should always be thinking of ways to expand the concept in any way possible.

The title of the song is an important indicator of the concept. It can simply act as a reminder of the central concept for the band, but another function of the title relates to marketing. The title can be the hook that causes someone to buy the song. It helps if the title is memorable. It should be contemporary and fit the style of music. The title is a marketing hook, and it's much better when it supports the concept.

Homework

You should complete these steps before you go to the pre-production meeting. Take home your recordings of the songs and go through each of them. Do the following:

1   Figure out what the concept is. Listen to the song and see whether you can discern a concept at all. Normally there is one, although as we mentioned, the concept could be that there is no concept. If you discern that there is a concept, *write it down.*
2   Critique the quality of the concept. Figure out on a scale of 1–10 how deep the concept is compared to other concepts in the world. This can also be helpful later in determining which of the songs on the album has the most hit potential. Occasionally, little adjustments

in the lyrics or other nine aspects can make the song focus more clearly on a deeper concept.

3   Go through each of the other 12 aspects and see how much they support the concept. Make a checklist of each of the 12 aspects and rate each on a scale of 1–10 as to how well each aspect supports the concept. If you need to, write down why you gave each of the 12 aspects the rating you gave it. Of course, at this point you can't rate the density of the arrangement, mix, or equipment because those are yet to be determined. Often only the lyrics will clearly support the concept. Occasionally you will find that the other components do support the concept in some way. Occasionally, you will come across an aspect that is supporting the concept so well that it could become one your main hooks of the song. If this is the case, write it down. It is especially important to notice where a component is actually distracting from the concept. This is where the real work should be focused – weeding out the problem distractions.

For example, sometimes the lyrics will have a line or two (perhaps a whole section), that seem(s) to have no relation to the concept of the song. Perhaps the rhythm is not supportive of the feel that the concept elicits. Occasionally, the performance might not be in tune with the emotion of the concept. Look for these distractions in each of the other aspects.

4   The most important step is to try to think of things that add to the concept. Come up with new ideas to help bolster what the band is trying to get across in the song – ideas to fill in the concept, to make it stronger, more powerful, and simply cooler. Now that you have the concept clear in your mind, it is much easier to figure out ways to add to it.

Take the things you found in any of the other 12 aspects that support the concept, and for each one ask yourself, "What could be added to this aspect to make the concept stronger?" Listen for any new ideas that might pop into your head.

5   Critique the title of the song. See if you like it and if it fits. If not, try to come up with ideas for a different name. You don't necessarily need to come up with a new title yet; you can wait and see what the band thinks. However, do decide whether the title fits the song and whether it is a strong name that sounds like a hit song.

Approach in the Meeting

As we discussed in the "Homework" section for this aspect, there are three things that you want to bring up in relation to the concept:

- The concept and ways to make it stronger
- Possible ideas for marketing the song
- The name of the song.

In the meeting, announce that you are going to discuss the concept. Tell the band that you have listened to the song and you came up with what you think its concept is, but you want to hear their ideas. After all, they are the ones who came up with the song and all the musical parts. Tell the band what you think the concept is, and then have everyone in the band say what he or she thinks the concept of the song is. Sometimes the guy who wrote the song will say, "Well, when I wrote the song, the concept was this. Then Joe wrote that stupid guitar part and completely screwed up the song." And then Joe says, "Screw you, man." The band breaks up, and it's over. Go home. You're done. Next band.

Or, you pull out your referee hat and say, "Guys, it's not worth it."

As each person in the band explains his or her impression of the concept, some will inevitably say something like, "I don't know what the concept is." That's okay. Even if people have differing views of what the concept is, the idea is to get all these thoughts out on the table. Even if no one agrees, you know everyone's opinion and how they might all fit together (or not).

More commonly, what happens is that you come to a consensus as to what the song is about. From then on, if anyone comes up with an idea that doesn't fit the concept, you can easily say, "No! Remember, we agreed the concept is this . . . It doesn't fit." More importantly, once you have the concept down, you have a context within which to be creative. You know the landscape in which to play. Throughout the rest of the production process and the recording session, you have a guiding light on which to base all critiques and judgments of any creative ideas. Everyone is now clear on what the goal of the song is.

Next, enlist everyone to try to come up with ideas within each of the 13 aspects to strengthen the concept. Throughout the rest of the project, you can use the concept to come up with ideas within each of the 13 aspects. Once you know what the basic concept is, you can be on the lookout for ideas that fit.

If it's appropriate, bring up any ideas you might have come up with for tying the song to external marketing projects, such as commercials or events. Again, give the group homework to try to come up with ideas themselves. Finally, discuss the name of the song, whether it fits, and whether it is appropriate for the market.

Aspect 2: Intention

Intention involves holding a certain energy or emotion throughout the recording and mixing process *with 100 percent focus!*

We would assume for most of you reading this (even those who have been around awhile in the business) that intention might strike you as unusual or curious. However, it is has become the most import aspect in the music that David Gibson has been doing over the last 15 years.

The intention of a song is very similar to the concept of the song, but much simpler. For example, maybe the concept of a song is about a love relationship, however the intention might be simply, "love." Then you would hold the energy of love throughout the song. Not only would you as the recording engineer hold that energy, but you would help the band to hold that energy throughout the whole song.

Although, this may sound a bit "new agey," there is good scientific evidence to show that intention when held with 100 percent focus gets embedded in the music and carried through to the listener.

But even more importantly, it affects the way the musicians play. For example, if a band is enveloped in the energy of love while recording (as well as the engineer), then they will certainly play better. It often helps musicians to get "out of their head" and focus more on the energy of the song. Besides the key aspect of great music is music that powerfully evokes an emotion. The more people are in a particular emotion, the more it is conveyed to the listener.

This technique make not work so well if the song is about hate (or maybe it would . . . God forbid).

Meanwhile, if the intention is about something as powerful as "Universal Love," or "Connecting to Source Energy," then a whole other level of creative energy often ensues.

Again, this also means holding that energy when performing the mix.

Ultimately, when a band even holds an intention with 100 percent focus when writing the song, it is even better.

Whether your song or composition suggests a journey through time, everyday life and love, or high consciousness and healing, your intended outcome is predicated on the basis of your mindset and your choices. You the individual are the instrument and everything at your disposal be it clarinet, piano, guitar, synth, or voice are simply tools used to activate and communicate your intention. Once you recognize that you are at the center of it all and the tools that you use are only extensons of the musician.

David has a song called "Unconditional Love," and he was very careful to first invoke the energy of unconditional love before he started writing the song. He then noticed his breath as he meditated. He then set the metronome in the digital audio workstation program to the same tempo as my breath. He was careful to stay in that energy while writing the song and recording each part. When I had other musicians come in and play on the song, we would first do a little meditation and bring in the energy of unconditional love. Even while mixing we all held that energy. That song is now packed with positive intention and is much more likely to invoke the energy of unconditional love when listened to.

Try it and see what you think. May not be appropriate for every band or group.

Structure and Values
Normally, most bands won't have established an intention so there is nothing to critique.

Homework
See if you can come up with a short intention based on the concept. Practice meditating on this intention yourself.

Approach in the Meeting
Tell the band about the concept of holding an intention with 100 percent focus and see how they respond. If they think you are completely crazy, let it go. It they are interested let them know the intention that you came up with in your homework. Then, have them practice the song while holding the intention. And, make sure you hold the intention with 100 percent focus the entire time they are performing the song. (Later you will also get the recording engineer to hold the intention.)

Either it will fly or not. If it doesn't fly we guarantee it will bring a whole new level of positive energy to the recording, which will be directly transmitted to the listeners. They won't know what hit them.

Aspect 3: Melody

The *melody* is one of the central themes around which a song is based. Sometimes it is *the* central theme. Otherwise, commonly the central theme is the lyrics, the rhythm, or a great instrumental or vocal performance.

Structure and Values

The question to ask is, how important is the melody to the overall song? At what level does it operate? It is all based on the motive for the song. Are you writing from your soul to evoke an emotion, or are you writing to create a standard, classic love song? A melody that comes from your soul just comes. It's hard to edit your soul; you don't have much choice other than to trust it. On the other hand, if you are writing a pop song, you might create a melody that is going to be more catchy and hummable. The truth is, the level of importance that the melody commands is totally based on your feelings. Does it evoke enough emotion or intensity to be the primary component? The key here is to simply ask yourself, how much does it float your boat? You get the point.

The melody should also fit the song and support the concept. You simply gauge whether the emotion that the melody evokes is in line with the overall emotion of the song. Melodies span the full gamut of emotions – sweet, loving, sad, happy, angry, sorrowful, excited, and so on. Figure out the emotion that the melody seems to be giving you and check to see whether it is appropriate for the song.

Also, the complexity of the melody should fit the song and the other 12 aspects. A simple melody might fit a song with a simple rhythm and structure. But more commonly, a simple melody may be even more appropriate for a very busy mix with a busy rhythm. Likewise, a busy melody could fit a simple or complex song.

Melody Instrumentation

Once you decide whether the melody fits, it is important to pay attention to what instrument or instruments are being used to create it. The melody could be constructed by a combination of instruments. Also, it could be reinforced by other instruments. The instrument playing the melody can be recorded twice or it can doubled with another instrument sound. Choose the arrangement and instrumentation appropriately, depending on the level at which you want the melody to play. The melody and lyrics are often two of the most important guides to the overall arrangement of the song. If you listen to the rhythm of the melody, it can give you important clues for a rhythm that will work the best. Of course, the melody also dictates the harmony parts.

Melody Structures

The melody can also be reinforced with sub-melodies and counter-melodies. A *sub-melody* is a melody other than the main melodic theme; it can be played by any other instrument. For example, a bass guitar or guitar could be playing its very own melody in the background. In classical music, sub-melodies are very common. In a really strong song, you could strip away the main melody, and the melodies from the other instruments could carry the song. Obviously, the most important and coolest production value is when you have multiple melodies weaving in and out of each other in a very musical fashion. The individual melodies are often not as important as the way they interact with each other.

Counter-melodies either respond to or answer a main melody or are interwoven within the main melody. The counter-melody is an integral part of the melody. Rhythmically, counter-melodies are fractions or multiples of the rhythmical note values found in the melody. They function very well at creating more complexity – they can make a song more interesting and intriguing.

Another way of creating interest and smoothing out a transition is by adding a *lead-in*, or *pickup*, to set up a melody. A few notes before the beginning of the melody can function to ease in the melody or bring it in with a bit of fanfare.

It is also important to note how predictable or unpredictable the melody is. Predictable melodies are when you want people to be able to hum it and remember it. Most hit songs have predictable melodies that anybody can hum. On the other hand, songs with more complex creativity (more conscious music that deals with art, jazz, Pink Floyd, and even punk rock, to name a few examples) are designed to evoke a certain kind of emotion to get you to think. Their melodies are often unpredictable to keep you off balance so you pay more attention.

Occasionally, producers will consciously have a melodic structure change from predictable to unpredictable (or vice versa). A melody that migrates from predictable to unpredictable is commonly designed to make fun of predictable structures. For example, if you have song about Trump that starts out predictable, it might then become quite chaotic. Not only does it fit the nature of the man, but also it tends to become a parody. On the other hand, a song that changes from unpredictable to predictable is commonly designed to accent the structured melody.

Although it might take you a while before you can trust it, resolving to something that is beautifully consistent can be quite effective. Stevie Wonder did it in the song "Sir Duke" from the album *Songs in the Key of Life*. Just before the song begins, there is a short acid-rock guitar solo, and then the song begins with a very clearly structured melody. The transition highlights the structured melody, making it stand out. It is somewhat similar to the effect you get when a symphony is first tuning up, and then begins to play. It is the transition from chaotic to sweet structure. Take 6 and Bryan Ferry have actually used this effect in songs – beginning with the sound of an orchestra tuning up.

Melody Density

A melody line can be doubled (sang or played twice), or even tripled. Elton John once said he never had a hit until they tripled the acoustic guitar part. It is especially effective to have every instrument playing the melody at special moments. It gives you the symphonic effect. You can also double the melody with a different instrument. We'll discuss more about melodic density in the "Aspect 7: Density of Arrangement" section later in this chapter.

Melody Performance

The way a melody is sung can completely change its essence. Technically, this is often referred to as *phrasing*. There is a huge difference between a song that is sung straight and one that is sung with personality and soul. When you really get down to it, melody is just an avenue to bring the spirit through. The pure emotion and personality that come through are really the most important aspects.

Occasionally, a group (especially a jazz group) might improvise around the melody in the first chorus. It is sometimes helpful to establish the melody before you play around with it, so that the listener knows what the actual melody is.

Melody Mix

Simple melodies are quite often placed lower in the mix, with more effects. If the melody is simple, it can be followed even at a low volume, and there is more room for effects. A busy melody has fewer effects, is often a bit louder, and frequently has a brighter EQ to highlight the detail of the complexity.

As a producer, you need to be aware of copyright infringement. If a song has more than seven notes that are exactly the same, you could be setting yourself up for a potential lawsuit. As the producer you should keep an ear out for any melodies that seem to have already been used.

Homework

You should complete these steps before the pre-production meeting. Take home your recordings of the songs and go through each song. Perform the following tasks:

1   In your song map, note where the key melodies happen in the song. Note what instruments are playing the melodies.
2   Write down your first impressions of the melody. How does it strike you? How does it feel? What might you compare it to?
3   Evaluate how central the main melody is to the song. What is goal of the melody?
4   Ask yourself whether the melody fits the song. Is it appropriate? Does it support the concept of the song?
5   Consider whether the melody is too simple or complex (busy). Check it out and make sure you are happy with the way it is.
6   If you have a major melodic hook in the song, look for a place where it might be repeated. Or, think about placing it low in the mix in another section of the song. Or, you might use another instrument or part to hint at it.
7   Note whether there are any sub-melodies or counter-melodies. Could the melody be reinforced by sub-melodies and counter-melodies? Are there any places where you might add a lead-in?
8   Evaluate how predicable or unpredictable the melody is. Do you like it the way it is, and does it fit the song? Might it be appropriate to have the melody change from unpredictable to predictable (or vice versa)?
9   Ask yourself if the melody is being played or sung "straight" the first time it is introduced in the chorus. If not, be sure to make a note to ask the musician to give it to you straight the first time, so that everyone knows what the melody is up front.
10 Consider whether the melody reminds you of another song. Is it already used in another song precisely? If so, you need to bring up the possibility of copyright infringement with the band.

Approach in the Meeting

Listen to the song with the band, point out the main melodies, and discuss how strong they are. Present the ideas you have developed from your homework. Discuss each point with the band members. To save time, only discuss major points and components that might need refinement. Make judgment calls on what points to address. And be sure to get the band's ideas.

Aspect 4: Rhythm

Most commonly, we think about the rhythm of rhythm instruments, such as drums and percussion. It is also important to focus on the rhythm within each instrument, such as guitar, keys, or vocals, even if the musical part is primarily playing a melodic line. Even more importantly, you need to focus on the relationship of rhythms between all of the instruments in the mix.

Structure and Values

The type of rhythm that is used is based most commonly on the style of music. Each style has developed its own traditions. It is obviously different for big band, rock-and-roll, disco, reggae, hip-hop, and dance music (electronica). Some styles, such as rap, hip-hop, and dance music are still evolving; therefore, it is important to stay on top of trends.

Tempo

It is critical to pin down the precise tempo in the song so that you end up with exactly what you want when you go into the studio. Different tempos have different physiological effects on the body, but frequently the lyrics will dictate the tempo. Often the best tempo is whatever the songwriter came up with when he or she originally wrote the song.

Rhythm Density

Once a rhythm has been established, the most common critique is whether the rhythm is too busy or too simple. Anyone can tell whether the rhythm is boring or too complex. If you are falling asleep, the rhythm could probably use some spicing up. If you can't keep up, the rhythm might need to be weeded out a bit. Also, you might consider changing the complexity of the rhythm from section to section to add more variety and accent what is going on in the song. For example, you might suggest that the drum pattern be changed a bit for the lead break or bridge. A change in the guitar part might be appropriate for the choruses. Also, consider adding more complexity to the rhythm as the song progresses or as the emotional aspects of the song deepen.

It is also important to make sure all of the instruments create an overall rhythm that works. Make sure no two-rhythm parts are clashing. You might need to change the kick drum so it syncs up better with the bass guitar. Or, maybe you need to change the bass guitar or guitar pattern so they sync up together. One of the most common rhythm problems is that one or more rhythm parts are too busy.

The more basic you keep each instrument

the cleaner the arrangement comes out

The most common problem is a drummer who is playing too much. It is important with every instrument to leave room for the other parts.

Rhythm Instrumentation

Certain instruments have their own capabilities for creating different rhythms. Obviously, a tuba is not going to play a tabla or conga part. Also, certain types of musical parts, such as falls, stabs, and trills, are played best by certain instruments, such as trombones, trumpets, and flutes, respectively. Therefore, choices of instrumentation are often made based on the rhythmical component that is needed in the song.

A unique effect is to explore alternative instruments and sounds for a typical rhythm part, such as kick drum, snare, hi-hat, and so on. For example, you might have the congas play the hi-hat part. One unusual effect used in a few hit songs (such as "Layla" by Derek and the Dominos with Eric Clapton) is to switch the kick and snare sounds so the snare is playing the kick part on one and vice versa. In rap, it is common to use different instruments for the snare drum. You can use almost anything. In one of Stevie Wonder's R&B songs ("Ribbon in the Sky"), he uses a water drop from a faucet for the snare. There are a huge number of possibilities, including street sounds and nature sounds.

Counterpoint

One of the most interesting techniques is to use other instruments to fill in a rhythm. Take a rhythm part and divvy it up among multiple instruments. This can be especially effective when each of the rhythmical components is then panned to different places in the mix and effected differently with various studio effects.

Homework

You should complete these steps before you go to the pre-production meeting. Listen to the recordings of the songs and:

1   Check to see whether the type of rhythm fits the style of music. Simply ask yourself whether it feels good. Does it feel right?
2   Listen to see whether the instrument selection is right for the rhythm. Think about whether you might choose other instruments. Take copious notes.
3   Figure out the exact tempo in beats per minute (bpm). If you don't already have one, buy a metronome. (You also can use a computer sequencer if you have one.)
4   Critique the tempo. Listen to simply see whether it feels good and right to you. You can often tell whether the overall tempo is correct (or appropriate) by listening to whether the vocals sound rushed or too laid back.
5   Critique whether the rhythm is too busy or too simple. Look to see if the rhythm should get more complex or be more basic in any sections of the song.
6   See whether you would like to add any polyrhythms to spice up the rhythm. Note the accents and see whether it might be appropriate to change them. Consider adding pickup notes and grace notes.
7   Make sure no sound is walking on any other sound rhythmically. You should be able to hear both parts in your head at the same time and have them work together. Perhaps the guitar part is stepping on the keyboard part. Look for parts that are too busy.

8   Listen for any places where you might divvy up the rhythm among instruments. If so, think about how these instruments that are playing the different rhythmical components might be panned or effected differently.

9   Explore using different sounds for traditional parts, such as kick, snare, and hi-hat. See whether switching the kick and snare might be appropriate.

Approach in the Meeting

First, play the song and discuss any ideas for changing the rhythm to fit the style of music. If you have any misgivings about the feel of the rhythm, let the band know and see what they think. Point out anything that is too busy or too simple. If the band members agree that it needs to be changed, enlist the drummer or the appropriate band member to try to come up with a way to make the rhythm more simple or complex. If there are rhythm parts that are walking on each other, ask the band if they notice it. Then you might suggest, for example, that the guitar player and keyboard player get together and figure out the parts that work together.

Explain any ideas you might have about splitting up rhythm parts among different instruments. Give examples of different panning and effects and see what the band members think.

Get the band to explore with you the idea of adding more complexity or simplicity to the rhythm in different sections of the song. Even if you know nothing about the way rhythm parts work or fit together, you can get the band to do the work of fixing them. They know the songs better, anyway.

Tell the band the tempo that you detected, and let them know if you feel that it should be different. Ask them what they think. Make sure they are perfectly happy with it.

Aspect 5: Harmony

Harmony, at its most basic level, is simply two notes being sounded simultaneously. It is important to be aware of any inherent harmonies that exist between two sounds, whether produced by the same instrument or different instruments. Then, you look for places where harmonies could be created. For a lot of people, harmonies are one of the most important magical components.

Structure and Values

The most basic consideration is whether the harmonics complement or create tension with the melody. Occasionally, tension can be a good thing. Prince performed a clever study in the relationships of harmonies in the song "Controversy." Monk also used tension in the harmonies in the song, "Ugly Beauty."

There are primary considerations when it comes to creating harmony parts – placement, chord density, chord structure, chord instrumentation, harmony performance, and the way the harmony is performed and recorded.

Placement

Where are the harmonies placed in the song? Background vocal harmonies most commonly are added on choruses in a song. However, they also are placed as accents more sparingly throughout other sections of the song. In rap music they might even be referred to as *ad libs* (although these might simply be in unison). Some background vocals are simply another vocal part – not harmonies at all. They might be a response to a lead vocal, or they might be something like oohs or ahhs. These additional parts might even have their own harmonies.

Chord Structure

As discussed in the chapter on Music Theory, *chord structure* is what notes you use to create the chord, and the intervals between the notes. The type of chord you choose in any harmony should be based on the emotional feeling that you want. The most basic consideration is whether the chord is consonant (sounds harmonious), or is dissonant (creating tension). Although, there are many finer distinctions as discussed in the Music Theory chapter.

Chord Density

Chord density refers to the number of notes that make up the chord. Chords can be in unison (simple doubling of the part – sung or played twice), one part, two part, three part, and so on. The more notes in the chord, the more it will sound like parts of a symphony. Also, the more parts contributing to the chord, the more it will fill out the mix.

You can also have more than one person singing each part. Just to give you the range of possibilities, imagine a seven-note chord with seven people singing the tonic (bottom) note, six people singing the next note, and so forth until the top note is only sung by one person. With wide panning, you end up with a chord pyramid, which would look like what you see here:

Visual 78
Chord Pyramid

As you can imagine, the possibilities are endless.

Chord Instrumentation

A chord can also be spread across different instruments. You could have a sax play the tonic note, a tenor sax (or trombone) play the middle note, and a trumpet play the high note. It gets really interesting when you spread the chord across an unusual array of instruments – for example, bagpipes, sitar, and accordion. More commonly, you might have the strings play the upper notes of a chord. We have even had the vocals sing the 1, 3, 5, and 7 intervals, with horns playing the 9, 11, and 13 intervals for a really unique spread.

Harmony Performance

The tightness of the performance of the chord also makes a difference. For example, many of Steely Dan's songs have the vocals doubled so precisely that it is sometimes hard to tell whether they are real or doubled with a digital delay. In a lot of rap music, the looseness of the harmonies creates a much looser and more "real" feel, in a sense. Often it is nice to have a harmony part last a little longer than the original melody line.

You also can create an incredible arrangement by using other instruments to function as your harmonies. For example, you can have strings or a synth sound play the harmony to the lead vocal. Or, you could double the harmonies of a background vocal section with synth harmonies.

The Way Harmonies are Recorded and Mixed

There are a number of options as to how harmonies might be recorded and placed in the mix. Each note of the chord can be spread left to right in various ways. You could use two mics and record a three-part harmony in stereo so it is panned left to right, as you see here:

Visual 79
Three-Part Harmony Panned
Left to Right

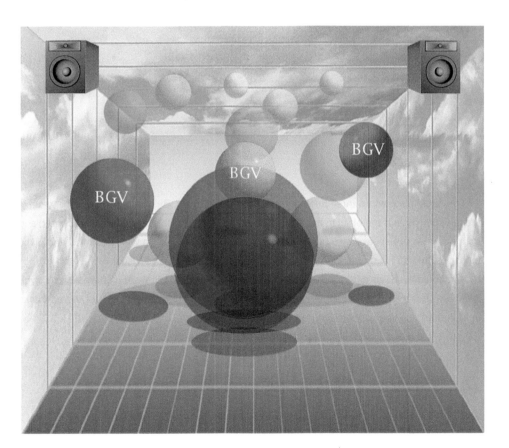

We often prefer to pan them so the high part is in the center. If you think of pitch as a function of height between the speakers, it forms a pyramid of sorts.

You could record a three-part harmony on one microphone and place it in the left speaker in the mix, and then record the same three parts on another track and place them in the right speaker, creating a full stereo spread of harmonies. (Two similar sounds like this will normally pull together, as shown below.)

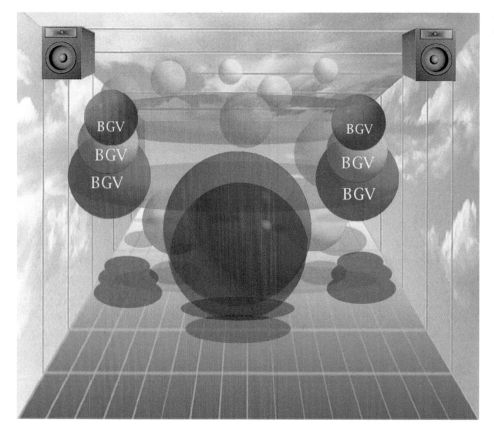

**Visual 80
Three-Part Harmonies on Two
Mics Panned Left to Right**

You could also record three people in stereo with two mics. By putting one mic (or track) in the left speaker and the other in the right speaker, you will hear the three parts left, center, and right between the speakers. Then, record the same three parts again with two mics and place them so that you have two parts of each left, center, and right. (See Visual 81.)

You could even mix and match at this point. You might pan the first set of stereo tracks (with the three vocals left, center, and right) left and right. Then, pan the second set of stereo tracks right and left, so you get the layout shown. (See Visual 82.)

Visual 81
Doubled Stereo Miking of Three
Vocals

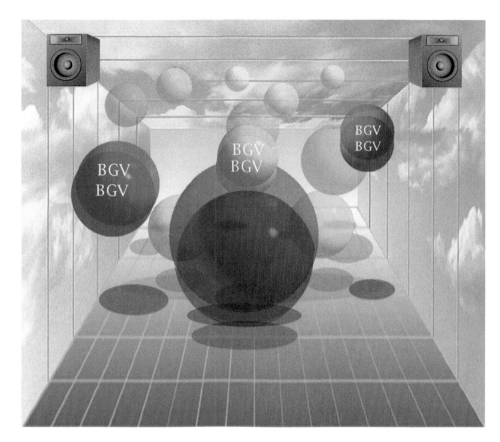

Visual 82
Mix and Match

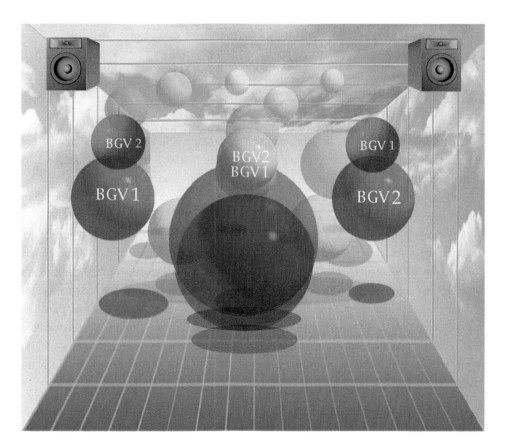

You also could record a three-part harmony 16 times on 16 different tracks so that you have 48 vocals. Then, ping-pong (bounce) the 16 parts down to two open tracks. Once you have mixed the 16 tracks down to only two tracks, you can erase the original 16 tracks and reuse them. You end up with 48 vocals on two tracks in full stereo for that huge Mormon Tabernacle Choir effect. (See Visual 83.)

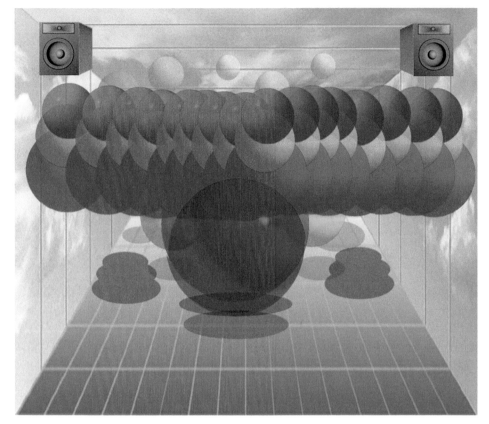

Visual 83
Mormon Tabernacle Choir
Fullness

Using some of the new digital technology, you can create a different vocal sound by adjusting the pitch and formants (harmonic structure) of the sound. That way, you can get multiple vocals that sound different out of one singer, creating a fuller background vocal section.

You normally want to have the band at least double the background vocals (record them twice), so you place them left and right, balanced between the speakers. (See Visual 84.)

For one part of the song you might use one way of recording the background vocals, and for another section you might use another way of recording them, therefore creating different textures for different parts of the song. For example, in the verses you might record the harmonies in unison and pan them to ten and two left and right between the speakers. In the chorus you might record three-part harmonies twice (double tracking them), and then pan them completely left and right. In the bridge section of the song you might record a two-part harmony with a unique chord structure with a flange on it and place it low in the mix. And finally, in the vamp at the end of the song you might record 16-part harmonies with strings doubling all the parts and pan them all completely left and right.

Visual 84
Double-Tracked Background
Vocals Left and Right

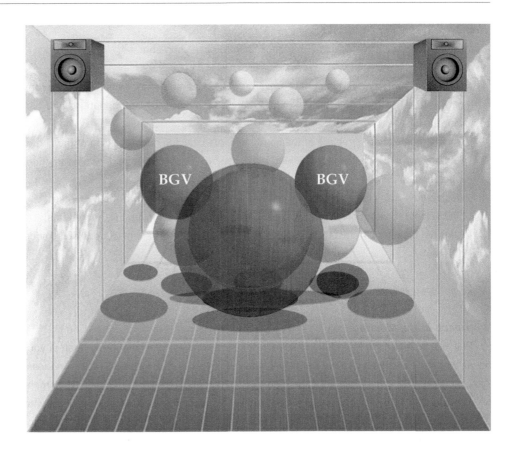

A common approach is to add more harmony parts whenever the song gets more intense or emotional. Any of the 13 aspects could contribute to building this emotional intensity. The most common example is created by a lyric or performance of a lyric. However, sometimes it is a combination of instruments playing off of each other. Harmonies might be thought of as "the icing" on the cake at peak moments. Since they are some of your most magical tools in production, they can be used to take things over the top into another level of emotional intensity.

There are many possibilities that musicians often don't think about. There are also harmony parts on instruments other than vocals. Often, a lead instrument might have a harmony included. A melody line on a synth might be harmonized. You might harmonize a trumpet and sax part. You might even harmonize a rhythm guitar part.

The overall consideration here is, do the parts actually add to the song, or do they distract from it? When making this judgment it is important to consider how the background part is mixed. If it is placed low in the mix with an effect, it will be perceived quite differently than if it were placed at the same volume as the lead vocal.

Finally, think about whether you want to spend time getting every harmony part perfect, or whether it would be acceptable to copy and paste harmony parts from one section to another (from one chorus to another chorus, for example). The song might not end up as rich (although some people prefer the consistency), but the time savings can be well worth it.

Homework

Before you go to the pre-production meeting, listen to each song and:

1   Note where the current harmonies are. On your lyric sheet, mark where each harmony part will be. Do they add to the song? Where else might it be appropriate to add harmonies?
2   Decide on the chord structure of the harmonies. Also, decide on the voicing of the chord.
3   Decide how many parts the harmonies will consist of – unison, one part, two parts, three parts, and so on.
4   Decide how many layers and how many tracks you will use to record the harmonies.
5   Decide how the harmonies will be mixed as far as volume, panning, EQ, and effects.
6   Consider creating different textures by treating harmonies in different sections of the song differently.

Make notes on your lyric sheet for all of these considerations.

Approach in the Meeting

Pull out copies of the lyric sheet with your notes on harmony placement. Share all of your ideas with the band and see what they think. Then get them to come up with their own ideas. You can work on it in the pre-production meeting, but you might also give them an assignment to work on it at home for the next meeting. Finally, discuss how picky you want to get with how a doubled part is performed. Point out that it takes much more time to get it perfect. Also, discuss whether you might want to copy and paste any harmony parts from one section to another to save time.

Hiring a Music Arranger

Do as much as possible with what you know to critique, fix, or expand the melody, harmony, and rhythm. But depending on your musical knowledge, you might want to take the music to a music arranger (who is similar to a music producer). This is someone who can listen to the recording and check out the musical parts.

Be sure to tell the band that they don't have to use any of the arranger's ideas if they don't like them. Normally, you can get an arranger to do a quick listen for only $50 per song. If they are going to do serious arranging, it might be $200 or more per song.

Check with your local recording studio to find a good arranger. Almost all major studios know of arrangers and producers that have worked there. Ask for someone that specializes in your style of music and will fit your price range. Hiring an arranger to go through the songs you are producing can make a huge difference.

Aspect 6: Lyrics

There are two main components to lyrics:

*   The words
*   The musicality, or lyricalness, of the words (in other words, melody and rhythm).

Structure and Values

The first component, the words, is practically sacred. If you are going to critique the grammar or the message, you might as well think of it as a song rewrite. Remember, there are a huge number of hit songs whose lyrics are grammatically incorrect and/or make no sense at all. There are tons of songs where they are just making sounds instead of using words. After all, it is art – poetry, so to speak. However, there are a few items to look out for when examining the words.

Sometimes lyrics support the concept of the song throughout and sometimes they don't. You've probably already noticed that there is a wide range of songs that vary in terms of how closely the lyrics support the concept. A large number of songs have lyrics that go off to who knows where. Of course, this is not against the law; in most (but not all) hit songs, every word has a direct, preconceived reason for its existence. In songs that tell a story (which are common in country music), every word supports the concept.

Likewise, the lyrics will sometimes match the progression of the song. In a song about making love, the song might climax right along with the lyrics. In a song about a breakup, the lyrics and song might also climax together (in a whole different emotional manner). The roller coaster of feelings and emotions that lyrics evoke should fit the music in some way. There is also the smoothness of the roller coaster. Do the emotions change dramatically, or do they have a smooth progression (perhaps slowly getting deeper and deeper)?

Some producers value a preservation of words. If words don't really add anything, perhaps they should be taken out. Many songs are just a bit wordy for no reason at all. Of course, words can only be taken out if it happens to work rhythmically.

Finally, because of copyright laws, it is often a good idea to keep an ear out for lyrics that resemble those in another song. Clichéd, banal, or stereotypical lyrics should be avoided. You can also make your own judgment call if you would like to comment on lyrics that are too violent, involve racial slurs or sexism, or are downright evil. We've been known to not allow back-masking (backwards recording) of evil messages.

When examining the musical content of lyrics, one of the most important things to watch out for is that the lyrics are rhythmically correct. It is critical that the lyrics fit the music rhythmically. This is important in all styles of music, but it is especially important in rap. Sometimes a rap artist will have too many words and say them really fast to get them in, even though they are rhythmically incorrect. For example, the line, "Get on down with the bad self," has just the right number of syllables and rolls right off the tongue, whereas, "Get on down with the really, really bad self," does not.

Occasionally, you could have lyrics that are intentionally out of rhythm, so that when you get back in rhythm, it's really cool.

There are also different types of lyrical phrasing. In layman's terms, *lyrical phrasing* can be thought of as how many notes are given to a word and how they are divided up rhythmically. Originally, Nat King Cole started out by giving each word its own note. He is considered the father of structured phrasing. Louis Armstrong's way of phrasing set the standard for traditional and current vocal and instrumental jazz phrasing. Stevie Wonder opened up a whole new world by singing an entire song in one word! He would put more notes into one word than should be allowed by law. Hank Williams Sr. set the standard for phrasing in country music. Thomas A. Dorsey was the father of gospel phrasing, and James Cleveland and Aretha Franklin took it to a whole other level. Jerry Lee Lewis, Little Richard, and Chuck Berry helped set the standard in rock-and-roll. Bob Dylan, James Taylor, Joni Mitchell, Bob Marley, Al Jarreau, Eric Clapton, Led Zeppelin, and even some death metal bands have all utilized very unique lyrical phrasings. Now, with the introduction of rap, a whole other world has been opened up. Tupac Shakur, Snoop Dogg, and Eminem have all come up with their own unique ways of rhythmical lyrical phrasing that are completely distinct.

Another consideration is the clarity of the enunciation of the words, commonly called *articulation*. You can make your own judgment about whether this is important. If the meaning of the words is the primary component of the song, then it goes without saying that the lyrics should be readily understandable. This also might mean placing them a bit louder in the mix. The clarity of the lyrics can become a very important issue when harmony parts are also mimicking the exact same words.

Songs are written in two ways – lyrics first or music first. When the lyrics are written first, they normally have their own rhythm and musicality. Most people who write poetry commonly have a melody and rhythm in their head as they are writing. This inherent musicality can be an excellent guide to what might go on in the music. As a producer, you can use the music inside the lyrics to come up with some musical components for the song that can be very powerful. Also, when the lyrics are written first, often the music tends to end up with more turns and changes. This is fine if it still flows.

For most of Elton John's songs, his songwriting partner, Bernie Taupin, would write the lyrics first, and then Elton would put music to them.

On the other hand, when the music is written first, the lyrics then have to fit the form of the music. Sometimes this constrains the ease with which words flow. Often the trickiest part of songwriting is getting the lyrics to fit the music, which can end with disastrous results: Either the lyrics get changed so they no longer get the message across, or they just don't fit musically.

Of course, when the two fit, it is magic. But what is really incredible is when they intertwine in a way that is magical (much like when different parts of a Bach fugue or a Beethoven symphony are intertwined). You should watch for such synchronicity of music and words and treat it with the utmost respect and delicacy.

Going out on a limb to help refine weak lyrics could mean the difference between a hit or simply an overall great song.

It is important to know that if you add one word to a song, you then own part of the song by law. As you can imagine, this makes some songwriters quite wary of taking any suggestions. Therefore, because of copyright laws, a producer should be extremely careful when making suggestions in this area. Instead of coming up with new lyrics yourself, it is always better to try to get the songwriter to come up with new ideas.

Homework

As previously covered, you should already have the lyrics down on paper before the meeting. Be sure to make copies for everyone in the meeting. Use your own copy to write down comments or issues.

1   Listen to the lyrics and compare each line to the concept on which you have already decided. See whether each line supports the concept or if it diverges too much. Remember, it might not even matter that the lyrics aren't supporting the concept. Perhaps there is something else about them that makes it okay.

2   Listen for a level of magic in the way the lyrics synchronize (or don't synchronize) with the music.

3. Listen to make sure the lyrics are rhythmically correct. Don't forget to make notes on your lyric sheet about where and what a problem is. As in every case, make a judgment call about how bad it is.

4   Check out the lyrical phrasing on every line and see if it does your heart good. Listen to determine whether there is too much singing on any single word, or if there could possibly be a "hot lick" added to a certain word. Make notes on the lyric sheet.

5   Listen for any enunciation (or articulation) problems that really bother you.

6   Ask yourself, "If this word or line in the lyrics were taken out, would the meaning be affected?" If so, see whether it is at all possible to take out those words.

7   Listen for any words that are clichéd, overtly offensive, or outright evil. Make your judgment as to whether you want to fight that battle with the artist.

Hot licks are short musical riffs that are often contain a lot of emotion. They are sometimes called "ornaments." They might be as simple as vibrato, or as complex as a quick sequence of notes on one word.

Approach in the Meeting

Be careful, because some songwriters are very touchy about *their* lyrics, and rightfully so. Not only is the song *their* creation, but they stand to lose rights to the song if anyone interferes. Because of this, it is sometimes best to work with the songwriter one-on-one. If not, you might get the rest of the band (including yourself) to sign a release whereby they agree to contribute any lyrics for free with no attachments or future lawsuits if the song becomes a hit. Make it clear to the group that you want to preserve the sanctity of the songwriter's lyrics. Sometimes the songwriter would still rather come up with all lyrics himself or herself. Honor this without question.

Hand out copies of the lyrics to everyone in the meeting and have the songwriter read the lyrics of the song out loud. If the songwriter would prefer not to, read the lyrics yourself.

First, bring up any lines that might infringe on another song's copyright. Then, explain how the lyrics should support the concept as directly as possible. Point out the lines that seem to diverge, if they are bad enough. Open it up for discussion. If you found, when you did your homework, that some words could be deleted, see what everyone else thinks of the idea. On items of lyrical phrasing you might have the singer and songwriter work together. This could be a homework item for them.

Most important, if you heard lyrics (when doing your homework) that have too many or too few syllables, point that out. Tap it out on your knee to demonstrate if necessary. See how much it bothers the band. If everyone agrees it needs refinement, get the songwriter to come up with something else. This could also be a homework assignment.

If the lyrics are synchronizing with the music in a magical way, point it out so that you can remember to tell the engineer to make sure this magic is highlighted.

Prepare yourself, and bring up any clichés or offensive material. You might save this battle for last.

Make sure that everyone is happy with every word.
You don't want to go into the studio and record lyrics
that someone is not happy with.

Aspect 7: Density of the Arrangement

The *density of the arrangement* is defined as the number of sounds in the song at any single moment, including how many sounds are in each frequency range. It is commonly also called *placement.* It is about knowing the range capability, dynamics, and where a particular instrument sounds and fits best in the harmonic scheme of the arrangement in order to best maximize its usage and intension.

Structure and Values

Because legato sounds last longer, they will create a denser mix. For example, strings take up a lot of space in a mix because they are normally legato. An occasional note here and there leaves plenty of space for other sounds and makes for a clearer mix. The more notes played, the busier the mix, but legato sounds take up the most space.

Although much more subtle, the complexity of a sound's waveform also will affect the fullness of a mix. For example, a flute, being practically a pure sound, will create much less density than a violin, which has many more harmonics present. A distorted guitar with all of its harmonics will fill out a mix much more than a harp. A piano not only has a large number of harmonics, it has three strings per note, making it even richer.

An instrument can also be double-tracked (the exact same part recorded twice) or even triple-tracked. Vocals are commonly double-tracked, especially when the singer is a little weak. Guitars are most commonly played twice. Miking an instrument in stereo with more than one mic also can create a fuller sound. On one project, we had a guitar split into two different amps with four mics on each amp, all mixed in stereo to two tracks. We then had the guitar player play his part three times! We ended up with one huge guitar sound. Adding time-based effects, such as delays, flanging, or reverb, also helps to fill out the arrangement because they actually create additional sounds.

You also could double a part with a different instrument. For example, you could have the keyboard play the exact same part as the guitar. With a MIDI guitar that sends out MIDI information to a sound module, you actually can get a guitar sound and any keyboard sound you want at the same time. When you are working with the computer, it becomes a no-brainer to add more sounds. With MIDI, all you have to do is send the MIDI signal to more than one MIDI channel, and you're playing multiple synth sounds at once. We'll discuss other possibilities in Chapter 10, covering digital audio production.

The overall density will create an emotional dynamic. (See Visual 85.)

A solo piano and vocal obviously will be perceived differently than the same piano and vocal in a full arrangement with 20 other instruments and a full orchestra behind them. (See Visual 86.)

Visual 85
Solo Piano and Vocal

Visual 86
Piano in an Orchestral
Arrangement

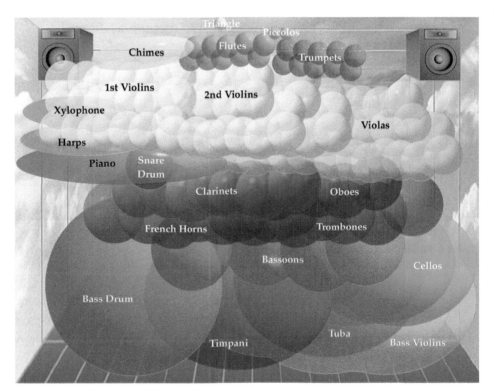

However, the most intense dynamic is when the density changes – that is, when sounds are added to make the mix fuller or removed to make it sparser. It is the transition that creates the most effective dynamics. For some songs, it is appropriate to have consistent density, such as in folk music and punk rock. (Amazing to think those two have something in common, huh?)

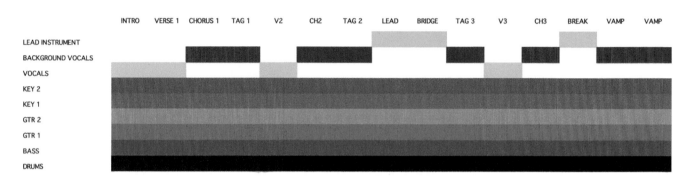

Chart 5 Consistent Density

In other music, the song calls for building and breaking down the density of the arrangement.

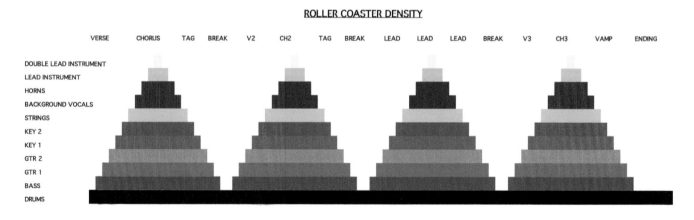

Chart 6 Roller Coaster Density

The smoothness with which the density changes is an important factor. Many people feel that density changes are most effective when they happen in stages (in other words, in small steps). Others feel it is much more effective to go from an extremely full arrangement to one sound in a single moment (breakdown or drop). (See Chart 7.)

BREAKDOWN DENSITY

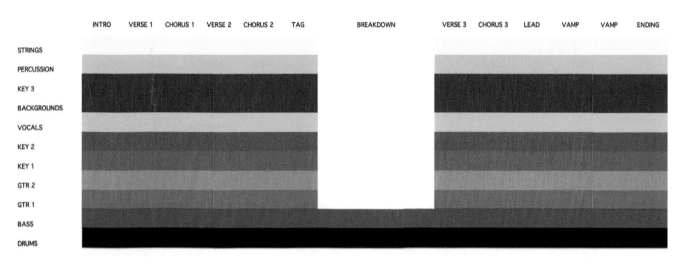

Chart 7 Breakdown Density

In dance, electronica, and techno, all types of transitions are utilized, sometimes in one song. In some songs the density will slowly build and then slowly break down like this:

PYRAMID DENSITY

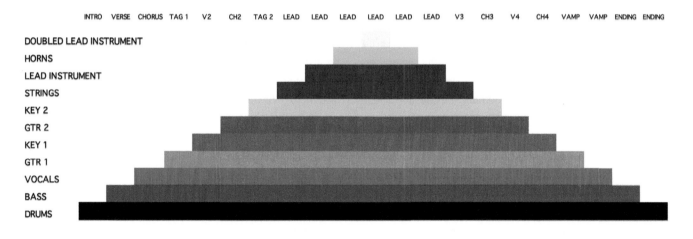

Chart 8 Pyramid Density

In traditional pop, about 75 percent of the time the density simply builds, becoming denser and denser from beginning to end like this:

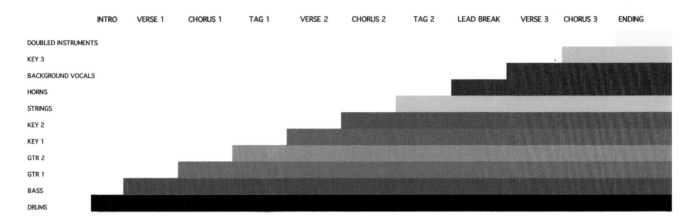

Chart 9 "Stairway to Heaven" Arrangement Density

Density is dictated by the dynamics of all the other aspects, but there are a couple of primary considerations – the emotion present in the song and the density that the songwriter came up with in the first place.

First, the number of sounds in the mix often follows the overall emotion of the song. As the emotion of the song builds, so does the density of the arrangement. As the song gets more personal, so might the arrangement. In certain cases the density dynamic can be flip-flopped so that the arrangement thins out as the emotions get more intense. Or, as the song gets more personal, the arrangement might get more intense.

As previously mentioned, it is especially important to pay attention to the number of instruments being used to create the melody. The melody can be constructed with a combination of instruments, and it can be reinforced by other instruments. The instrument playing the melody can be recorded twice or it can be doubled with another instrument sound.

When a band gives you an arrangement, the question is, does it need more or fewer sounds at any place in the song? (See Visuals 87, 88.)

If a band gives you a full arrangement there are two options: You can either make it even fuller or you can weed it out. Of course, it totally depends on what is appropriate for the particular section of the song. If the band obviously is trying to create as full a mix as possible, you can add more sounds, notes, or effects.

However, a more common problem with arrangements is when it is too full and needs weeding out. There are some bands that would record 48 tracks if available, simply because they can! Sometimes this can be a good production technique, as long as you then pick and choose which parts to use. In fact, in hip-hop and dance music, it is common to work with a large number of tracks that are brought in and out to create a dynamically changing arrangement.

Visual 87
Sparse Mix

Visual 88
Full Mix

On the other hand, if a band gives you a sparse arrangement, the question becomes, do you leave it sparse or add more sounds? Again, listen to what the song is telling you. Let the details of the song (the 13 aspects) dictate the answer.

Besides controlling the number of sounds in a mix, it is also important to think of the density of each frequency range. The most common value is to have sounds distributed equally among all the frequency ranges.

Visual 89
A Mix with a Good Frequency Spread

Normally, you don't want too many instruments cluttering up any frequency range. However, occasionally you can create clutter in a range just so it will feel good when it is cleared up later. Play with fullness and sparseness at each frequency range to create another level of dynamics. You can distribute sounds by simply choosing instruments that naturally play in that specific range. You can also have an instrument play exclusively in one frequency range, or you could even pitch-shift a sound to a different range. Also, you can move sounds up and down a little bit by using different chord inversions or by using a bit of equalization.

Some songs are not meant to have a lot dynamics. Perhaps you just want to show off the subtleties of the singer or song.

Set the density of each range the way you want it,
instead of the way it happens to fall.

Homework

Before you go to the pre-production meeting, pull out your song map and listen to each song.

1   Ask yourself whether there should be more or fewer sounds at any moment. Listen to see whether any sound might benefit from being double- or triple-tracked. Consider whether doubling a guitar part would help the song. Think about having a keyboard part double the guitar part. Think about adding parts, such as a percussion part, or think about having another guitar part playing off of whatever the first guitar is playing. Try to come up with as many different ideas as possible.

2   Should you have more or fewer notes, or more sustaining, legato sounds?

3   Should you have more effects filling in the mix?

4   Most important, figure out an overall progression from simple to complex (and vice versa) that fits the song – or no dynamics at all. You might add more instruments in the second chorus, even more in the third chorus, and a whole orchestra for the finale. On the other hand, you might take out instruments in the second verse to create more intimacy.

5   Take a look at the density within each frequency range. Is it cluttered in any range? Is it spread evenly? Is it the way you would like it?

6   This whole process will help you budget your tracks. After you have finished mapping all of your sounds into your song map, look to see whether you are going to have enough tracks. Consider whether you want to record each track in stereo on two tracks. If you decide that you don't have enough tracks, you might need to condense tracks. You might need to record the whole drum kit on only three tracks. You might need to record the bass on only one track even though you might have several signals coming off the bass. It is quite common to run out of tracks when you are recording a lot of things in stereo and doing a lot of overdubs.

Don't forget to make copies of your song maps for everyone in the pre-production meeting. Make copies of the original arrangement the band started with, but also make copies of the version that has all of your ideas added.

Approach in the Meeting

First, explain the use of fuller or sparser arrangements and why you think it would be appropriate to create a song with a consistent density (or rather, a song where the density changes).

Pull out your song map and show the band their original arrangement. Then, hand out your homework copy and explain your ideas on building up layers in some sections and taking out sounds in other sections.

Explain the overall progression, for example, from sparse to full to sparse. Then, explain each individual idea. Ask the band what they think and if they have any ideas to add. Most bands play live on stage and often don't think about taking out parts, because what would they do . . . just stand there? And often they don't think about adding parts, either. On the other hand, many band members have highly developed ideas about layering. Regardless, it is especially important to involve the band in this process. They will most likely be able to come up with very good parts because they are so familiar with the song.

It is critical that you get the band enlisted in figuring out the manner in which tracks are layered. Give them homework. What inevitably happens is, a week later, the guitar player comes to you and says, "I've got a really cool idea. I've come up with three guitar parts to play that all work together. And we can place one in the left speaker, one in the right, and one in the middle when we mix it." It's great when you get the band going. They often come up with the parts that fit the best.

Explain to them your ideas on spreading sounds evenly throughout the frequency range and get their ideas.

If there is a concern that you might run out of tracks, mention it to the band now. Go through the song map and see whether it reminds them of any parts that they might have considered adding. Otherwise, you could get into the session and the drummer might say, "Oh, I was really planning on laying down six cabasa tracks on this. There's room, isn't there? C'mon, we gotta have it. It's the thing that will make this song a hit. Six cabasa tracks." And you don't have enough tracks to do it.

Aspect 8: Instrumentation

The *instrumentation* is your choice of instruments and sounds, and the quality of those sounds. It can also be your choice of vocalists, including background vocals. This is often also called *treatment*.

Every aspect of a composition or song should support and tell the same story as the theme or title, regardless of whether the piece has lyrics or not. For example: a song or composition entitled, "We are love birds (in paradise)," might benefit by the use of piccolos and flutes in order to affectively illustrate and tell the musical story.

Structure and Values

There are three main considerations for instrumentation:

- Sounds that work well together
- Sounds that are of the highest quality, with no problems
- Sounds that fit the overall song.

Sounds that Work Well Together

Certain sounds blend naturally. For example, the classic blues instrumentation of drums, bass, guitar, piano, organ, and vocals has stood the test of time.

If two sounds have a similar harmonic structure, they will blend together in a mix; you can't keep them apart. Even if you pan them completely left and right, they will often pull together. This happens especially when you have two guitars or two sets of vocals. (See Visual 90.)

On the other hand, if the harmonic structure is dissimilar, the sounds tend to remain separate. Even if you pan a flute and a bassoon together in the center of the mix, to your ear they will always sound like two completely separate sounds.

Visual 90
Two Sounds Pulling Together

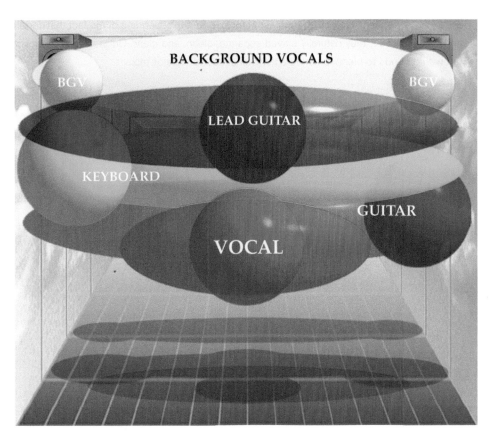

When you are deciding what sounds to include in an arrangement, this consideration couldn't be more important. Steely Dan was famous for their clean and clear mixes:

Visual 91
Steely Dan Mix

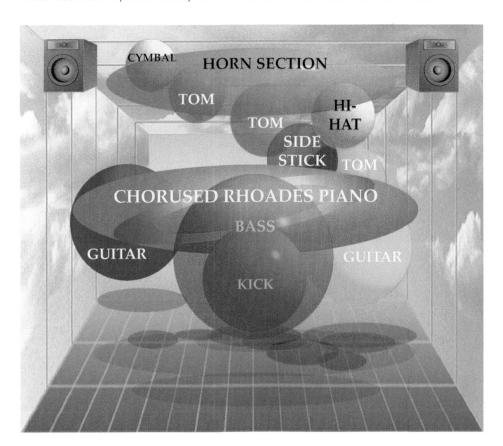

However, the truth is that the arrangements took complete advantage of the fact that certain sounds don't blend – they naturally stayed separate, clean, and clear. On the other hand, you might *want* sounds to blend, creating a full wall-of-sound style of mix.

Visual 92
Wall-of-sound mix

Sometimes we like the best of both worlds (when appropriate for the song), whereby you sometimes combine the wall-of-sound arrangement with some clean and clear sounds out front. Peter Gabriel does this in many of his songs.

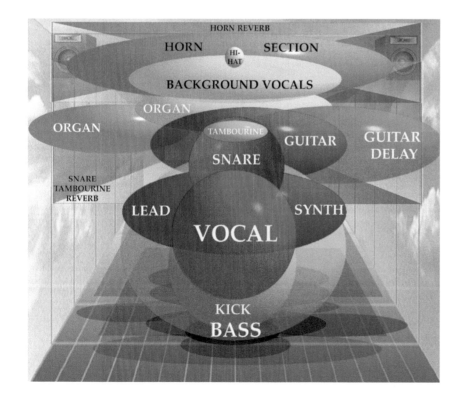

Visual 93
Peter Gabriel Mix

You also can go from one type of arrangement to the other within different sections of the song.

As previously mentioned, it is also important to make sure that the frequency range of your sounds is spread evenly throughout the frequency spectrum.

Visual 94
Mix of an Orchestra

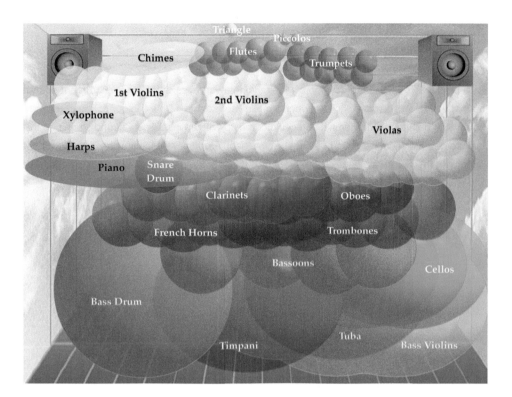

As the producer, you want to make sure that every instrument sounds good before you get to the studio. This means you want to be sure of three things.

1   There is nothing wrong with the instrument that would cause a problem with the sound.
2   You have a good-quality instrument that is putting out a quality sound.
3   You have sound that fits.

Problem Sounds

Problems include extra noises and buzzes, or just odd tonality. On drums, make sure all the heads are new. Also make sure that there is nothing wrong with the guitar sounds. Each guitar should be set up so that the intonation is correct. This means that both the nut and the bridge (which hold the strings up at each end) are adjusted correctly and that all the frets are at the perfect spot on the neck of the guitar. Otherwise, the guitar can easily be out of tune when you are playing higher up on the neck, even after you tune it with a tuner. If this is the case, get the guitar to a repair shop and get it fixed, or get another guitar!

It also good for all guitars to have new strings . . . but they must be put on at least a week before the session as they tend to stretch when first strung. New strings create more beautiful high harmonics.

Also, listen for buzzes in amps that you can't get rid of. Sometimes this can also be the guitar's fault – faulty wiring. Check every instrument that will be used in the session for hidden problems. The best procedure is to simply ask the musician if he or she knows of any little problems with the instrument.

These days, some software plug-ins can take the buzzes, hums, and pops right out of an instrument's sound.

Quality Sounds

As a producer, you need to develop a wide-ranging awareness of different instrument sounds so that you know the difference between a quality sound and a "low-fi" sound. In some songs, you would be hitting a homerun if you could come up with an instrument sound so unique that it becomes a "signature sound." At that point, people actually associate the memory of the song with the particular sound. Thomas Dolby did it with the bass synth sound in "She Blinded Me with Science."

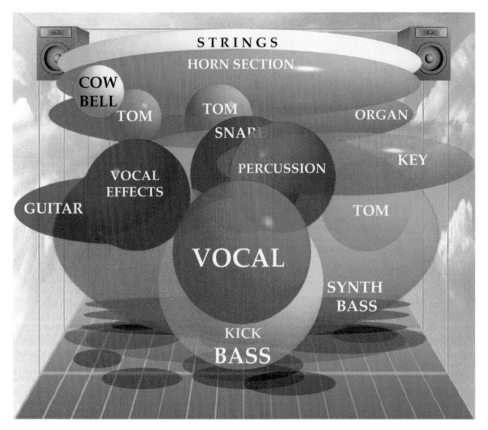

Visual 95
Thomas Dolby, "She Blinded Me with Science"

On the other hand, sometimes you can draw on the use of signature instrumentation. This refers to sounds or combinations of sounds that trigger a memory of another song, group, or musical style. Examples include the horn sections from Earth, Wind & Fire and Tower of Power. Other examples include bagpipes, didgeridoos, Tibetan bowls, crystal bowls and even timpani.

These sounds have their own associations and stigma (good and bad) that go with them. The Moog Bass sound and the 808 Rap Boom Kick are also good examples.

It's a good idea to learn the difference between different brands and types of instruments. You can study different instruments in books, but to really get to know them you need to hear them. One way to do this is to hang out at your local music store. Another way is to pay extremely close attention to instrument sounds that you hear in concerts, and in the studio. Check out the particular brand and model of the instrument making the sound. Of course, learning to play various instruments is helpful. If you come across an instrument for which you are unfamiliar with what a quality sound is, ask the musician what he or she thinks of the sound. You often can gain some insightful information this way.

If you have a drum kit that doesn't sound so great, see whether you can rent another set. There is no reason for a guitar player to use the one guitar he or she has for the entire album. Consider borrowing or renting some guitars so that the guitar player has a nice assortment of guitar sounds from which to choose. Normally, the album will sound much better with a variety of guitar textures.

Make sure all the amps sound good. Again, there is no reason for you to have only one amp sound on an album. It is especially effective to use a Y-Cable so one guitar can feed two different amps. Recording the two amps on two different tracks gives you a variety of sound combinations, creating a totally unique sound. You can even develop your own personal signature sound this way.

You should also become familiar with as many synthesizer sound modules as possible, so that you have a working knowledge of all the possibilities. A producer commonly will suggest unique sounds to incorporate into a song. There are a huge number of totally unique and bizarre sounds that you can use. And, there is a world of different ethnic percussion instruments available these days.

Often it is a good idea to design an altogether new sound in the synth or computer. Think about sampling some unusual natural sounds and using them as instrument sounds. When placed low in the mix, some very unusual sounds can actually fit in quite well, even in the most normal type of song.

We once recorded a heavy metal band that wanted the sound of chains. At first, we thought about going to the library to see if we could find the sound of chains on some sound effects CD. But then we realized that we had snow chains in the car. We went and got them, miked them in stereo, and slammed them on the floor for the song, "We Are Metal Prisoners." On the same song, the band also wanted the sound of a jail cell door slamming. We actually went to the local jail in this small town and asked them if we could sample the sound. They let us into the women's cells (because there were no women incarcerated), and we slammed the hell out of the doors. It sounded great with some large reverb. Be creative!

Sounds That Fit

The instrumentation should fit the style of music and the song. Each style of music has developed its own industry standard about what is cool at the moment. For example, it would not be marketable to choose an '80s drum sound and pattern in the new millennium. You should select the proper instruments based on industry standards – therefore, you need to stay in touch with musical styles and trends.

It is also important to choose your instrumentation based on what is actually going on in the song (the other 12 aspects). If it's a ballad, use mellower-sounding instruments. If the melody is intense, you might use an instrument with an edge. The rhythm might dictate specific sounds, and the lyrics easily could make you think of various instruments. The song structure can show you where the instrument sounds should change. If you have a dense mix, you might use sounds that blend (or not). If you have an incredible performance on a particular instrument, you might make sure that the other sounds are not competing by being too unique. Overall, listen to what the rest of the 12 aspects are telling you to do.

Homework

Prior to the meeting, listen to the recordings of the songs and:

1   Think about common instrumentations for the style of music you are working with, and decide whether you would like to subscribe to them. If so, think about where you might replace sounds and with what instruments.

2   Do the sounds fit the overall song and its concept? Rate it on a scale of one to ten. Go through each sound and critique its quality on a scale of one to ten. If it falls below a five, either get rid of it or figure out a way to spice it up with effects.

3   Do the sounds work well together? Conceptualize the overall blend or separateness of sounds throughout the song. Listen to make sure sounds are blending (or not) when you want them to. If they aren't, consider getting rid of or replacing the sounds.

4   Ask yourself whether the sounds are of the highest quality and coolest sounds available, with no problems.

5   Try to come up with a signature sound. Also, come up with unique sounds that you could place low in the mix. Think about including some unusual sounds that you can place in the background, because that's how you can create depth as a producer. When you are using percussion sounds, research to find some unique sounds. There are so many really cool percussion sounds available. There are hundreds CDs of unique sounds you can buy at music stores. You can also find unique sounds on the Web – just search for music libraries or sound effects CDs.

6   Think about whether it would be appropriate to have the band create its own sounds in the synthesizer and/or find its own samples out in the world. Think about possibilities.

Approach in the Meeting

At a rehearsal, check every instrument to make sure it is in good shape. Make sure the drums have new heads. Make sure the guitars have new strings at least a week before you record. When you first put on new strings, they go out of tune very easily. It takes about a week for the strings to stretch out. In a pinch, boil them.

Check out all the keyboards to make sure the sound modules are up to date. Keyboards are continually being updated with more complex and unique sounds.

Even if you know nothing about drums, say to the drummer, "Hey man, is this drum kit happening or what?" If he says, "No, it's a piece of crap," then go rent a drum kit (or at least a kick and snare) for the session. There is huge array of places that rent instruments. You can find them on the internet by doing a search for "studio instrument rentals."

In the pre-production meeting, point out any considerations you came up with as to how well the sounds work together. Discuss with the band your ideas on whether to have a normal instrumentation or one that is totally unique. Discuss the pros and cons.

Talk with the band about how to get better-quality sound. Point out any sounds that you feel are below fives on the scale of good quality.

Address the keyboard player and ask him whether he knows how to design sounds in the keyboard or on the computer. Give him homework to learn how to program his keyboard. Ask him to create a totally new sound. It will be really cool when he comes back the next week having learned how to program, and he comes up with an award-winning sound.

Talk with the guitar player to find out what guitars and amps he has. Ask him to bring in multiple amps to try out different sounds and combinations. See if anyone knows someone who will lend (or rent cheaply) a top-quality guitar and amp. Amps are much easier to borrow than guitars.

Talk to the band members about getting them to do their own samples. We've had people simply walk down a city street and sample street sounds and street people. It's amazing what you can come up with. It can also sometimes be very effective to walk around and just start banging on things. (If you haven't seen it, rent the video *Stomp* to open your mind about percussion sounds.) You never know where you might find the next hit kick drum or snare sound! Pay attention to the sounds everywhere in case you can use them. And don't forget that there is the whole world of nature sounds.

**Bring up in the meeting that
it would be nice to have at least
one signature sound on every song.**

Give the band a homework assignment to find at least one unique sound for each song. You don't have to use them, but often you get lucky and come up with a sound that can make the song. Remember: A sound that might seem really bizarre can become just the thing when it is placed low in the mix.

Finally, talk to the band about whether the sounds really fit the style of music and the details of the song. If refinement is required, do it.

Aspect 9: Song Structure

The *song structure* refers to the order and length of the song sections (intro, verse, chorus, lead break, bridge, vamp).

Structure and Values

Different sections of the songs are:

- **Intro.** Can be taken from any section of the song.
- **Verse.** The part of the song that establishes and tells the story. Also referred to as the *A section*. It normally supports the main idea in the chorus.
- **Chorus.** The main repeated concept of the song. It is commonly the main point or culmination of what was learned from the story of the verse. It is sometimes called the *refrain*, *hook*, or *B section*. Verse plus chorus equals AB form.

- **Bridge.** A section of the song that is quite different from the rest of the song. It is often used to give a different perspective to the homeland of the verse and chorus. The bridge sets up the climax, driving it home. It often sums up the song and emphasizes what the song is talking about. Sometimes it is used just to make the song more interesting.
- **Lead break.** This is normally the verse with a lead instrument on top. It is the most interesting when it is created as a bridge. In jazz music, the lead break often goes through a whole round of the song – verse, chorus, and bridge!
- **Vamp.** The chorus repeated over and over at the end of the song.
- **Tag or turnaround.** A short musical piece (normally only one to four measures) that is played in between sections of a song or even at the end of the song.
- **Breakdown.** A section where most of the instruments drop out.

> There is a theory that the basic blues structure is described in the "The Hero with a Thousand Faces" by Joseph Campbell. George Lucas also used this mythical structure in the *Star Wars* episodes. It has been shown to be prevalent in religion, fairytales, and myths throughout every culture in the world. It is so common that it is eerie. It begins with the hero at home in the homeland (verse and chorus), who then takes a journey into another land (the bridge). In this other land he often encounters some conflict or evil, which he overcomes. He then returns home (back to the verse and chorus) with some increased power or wisdom. Often this is just the new perspective gained by seeing another world. The homeland of the verse and chorus now has a new look with this perspective. It is archetypal.

Different classic song structures are covered in detail in Chapter 14, "Songwriting." The classic structure is the ABAB form: verse, chorus, verse, chorus. You can use any term to refer to a section of a song, as long as everyone agrees on the terms. We've given sections names as obscure as "The Squirrelly Synth Section," or as basic as "The Guitar Solo Section."

The most common adjustment that a producer will make is to the length of the song's sections. A five-minute lead break is obviously too long, but sometimes a verse and a chorus might be too long. For a song to get airplay, it generally needs to move along. However, this is not always the case, so make your own judgment call. Also, a record company (or your mother) might not be patient enough to listen to an introduction that is longer than ten seconds. It's practically against the law to have too long an intro unless you already have a hit. Often a producer will suggest doing two different versions, with a long and a short intro. You could also do a long version and edit the final mix to make a shorter version.

An intro can be thought of as a bridge between silence and the beginning of the song. Intros commonly include the main melody and/or chorus. Sometimes you simply use remnants or hints of other sections. Occasionally the intro takes you somewhere else completely, to keep the song interesting.

A producer also can suggest adding sections to a song, such as a bridge or a vamp. It is common to place a tag or turnaround at the end of a chorus before it goes back into the verse to let the song breathe a little bit.

One of a producer's most important functions is to help smooth over transitions from one section of a song to the next. Commonly, a song will be written with an abrupt transition from section to section – as if the sections are almost two different songs. There are numerous ways that transitions can be bridged. First, you can simply have another instrument play over the transition. It is also effective to have an instrument play a lead-in or pickup that overlaps the transition. One of our favorites is to have an instrument such as a sax or synth play an extremely long sustain note that is faded in slowly and then ends up being a pickup for a lead solo. It is quite common to use a tom roll to lead into the next section, but it is especially effective to use an extended tom roll as the intro to the next section. Jackson Browne once used a tom roll that faded in extremely slowly over an entire bridge section before the song broke into the lead break.

Homework

In each song critique the following:

1 Listen to see whether you need to add, take out, or shorten a section.
2 Ask yourself whether any section doesn't feel right. If it doesn't, figure out why and see what you can do to fix it.
3 Does every section of the song flow smoothly into the next section? Think of different ways to smooth over transitions, if necessary.

Make notes on your song map.

Approach in the Meeting

Pull out the map of the song structure that you created and hand out copies to everyone. First, ask the band if they are happy with the song structure. Often, this gives a band member a long-awaited opportunity to bring up concerns or ideas he or she might have been holding in.

Talk about the pros and cons of the song structure. Present your ideas for changes to the band. One band we produced had this song structure of verse, chorus, verse, chorus. Finally, we got up the nerve to suggest that the songwriter put in a bridge. "A bridge?" he said. "This song came from God." We said, "God likes bridges."

Don't forget to point out any ideas for smoothing over transitions.

Aspect 10: Performance

In this section, we're really only talking about preparing for a performance. We'll discuss actually critiquing a studio performance in Chapter 9, "Producing in the Studio."

Structure and Values

There are two types of performers. They can be categorized by comparing a jazz pianist to a classical pianist. A jazz pianist is always learning. This is not to say that a classical pianist is *not* learning new parts, but a jazz pianist seeks out new ways to be creative. It is important to realize what type of musician you are working with. A punk musician will approach a performance quite differently than a rap musician will.

As a producer, it is important to learn what a great performance entails. In fact, you probably wouldn't be reading this book if you didn't already have some pretty detailed values when it comes to a musician's performance. There are six main areas that contribute to a good performance:

- Pitch
- Timing
- Technique
- Dynamics
- Style
- Greatness (the goose-bump factor).

Pitch

Normally, the recording engineer is ultimately responsible for making sure all instruments are in tune and every note of a performance is in pitch. However, it is always helpful if the producer can lend a good set of ears as well.

There are three levels of pitch perception. *Perfect pitch* is when you can recognize the exact note or frequency of a sound. Some people can tell you the frequency (such as 440 Hz) when they hear a note. This skill, although great to possess, is fairly rare.

Absolute pitch is when you can tell what note a sound is. This is much easier because you don't have to be as precise as you do with frequency.

Relative pitch, the ability to tell whether a sound is in tune with previous sounds in a song, is much more common and extremely important. Although some people are born with the ability to detect relative pitch, it is also a skill that can be learned. There are some very good computer programs and study courses on tape that teach this skill.

We know from personal experience that almost everyone interested in this business can have very good pitch if they are able to hear the sound for long enough. So really, the whole key is to get quick at listening. You need to develop the discipline to ask yourself whether the sound is in pitch at *every single moment*. It all comes down to simply concentrating on finer and finer moments. You need to be able to hear the relative pitch of each note at the beginning, middle, and end of the note. As your concentration gets better, you can hear the pitch of every single note in a string of notes at a fast tempo. Of course, this amount of concentration becomes easier with practice.

The hard part is getting to the point where you can remember which note is out of tune in an entire riff. It is great if you can also tell whether a note is flat or sharp, but it is not absolutely necessary. It is enough to know simply if a note is out of pitch and which one it is.

Learning to tell whether a sound is in tune requires the discipline to focus on every single moment. The good news is that over time, it gets inside your body. Now whenever we hear a sound that's out of tune (even if it is in the background), it causes us to shiver or grimace.

Timing

You should be able to tell whether the timing is stable. Some people are born with perfect timing perception, but most of us have to listen very closely. Some people pat their leg. Others keep time with one finger in the air or by bopping their heads. Some just sit with their backs straight and tilt their heads in a funny way. Regardless of the technique, it takes serious concentration to hear variations in tempo.

If the drummer has experience playing with a click track, use one. Normally, it takes months of practice before a drummer is comfortable with a click track, where he doesn't end up chasing it – speeding up or slowing down, trying to stay in time. It takes even longer for a drummer to learn to play a "feel" on top of the click track. If you have at least a month before your session, you might have the drummer start playing to a click track.

Some people are fanatics about timing, and for certain styles of music and certain songs, this can be totally appropriate. The Eagles actually recorded a song multiple times and then compared each section of the song to a metronome. They would then choose the sections that they liked the best and splice them together – on two-inch tape! Now, of course, this would be much easier with a digital audio workstation. Also, as we'll discuss later, in some programs you can even quantize both MIDI and digital audio to the feel of a professional musician.

Listen for whether the drummer is consistently on the beat. Often, drummers will speed up when playing a tom-tom roll or when they are nearing the end of the song. Of course, sometimes a drummer will intentionally play on top of the beat or behind the beat to create a certain feel. Also, watch out for someone who is coming in too early or too late on a vocal part. With rhythm guitarists, listen closely to their consistency in timing. Make sure they are not rushing the attack of a note.

It is important to make sure that everyone comes in at the same moment when multiple instruments are playing at the same time. This is especially critical on the first and last notes of the song and whenever there is a stab or any staccato notes. Again, this can be adjusted in the computer if each instrument is isolated on its own track.

Technique

There are specific techniques that musicians must learn for playing any particular instrument; these vary depending on the style of music being played. Any tips or techniques you can offer to a musician can only help. Of course, you can't be expected to know the right thing to say to a musician for each and every instrument, but the more you work in the business, the more tricks you will pick up.

For example, there are specific techniques for playing each of the drums in a drum set. The kick drum should be "popped" with the foot. For some styles of music, it is best to really whack the snare drum. For guitar players, there are many little things to watch out for, such as not causing any string buzz from hitting the frets or not causing the strings to squeak as you move up and down the neck. All that is necessary is to point out the problems.

There are many comments that you can make to help singers (as well as a wide range of comments that don't help). Suggestions such as "Sing out" or "Project more" can be helpful if they are given at the right moment and with sensitivity. It's often a good idea for the singer to focus on using his or her diaphragm more often. Some people will bring in a vocal coach during the recording session to really help out.

It is especially important to pay close attention when anyone else in the room makes a comment or suggestion that works. After a while, you will learn an entire range of useful techniques to help musicians play better.

Dynamics

There are two main types of dynamics that you can critique and help refine. First, it is a good idea to keep an eye on simple changes in volume dynamics in a performance. You might find them to be too dynamic when they vary too much. Or you might find them to be too stable in volume, so that it sounds like a synthesizer or drum machine. It is important to make sure that the volume dynamics fluctuate in a way that is natural-sounding or appropriate for the song.

The second dynamic to critique is the level of emotional intensity at every moment in the song. Just as with volume dynamics, you might find an instrument's dynamics to be too varied, too boring, or just inappropriate. For example, singers occasionally sing too intensely at the beginning of a song when perhaps they should save their intensity for the end. On the other hand, maybe they need to put more emotion and feeling into the performance up front.

Listening to the performance dynamics at each moment in the song can help you fashion the song exactly the way you want.

Style

**There are singers
and then there are stylists.**

A *style* is a musical personality. It's a person's own personal hook. It is a collection of nuances that define the style – a recognizable pattern that grabs you in some way. It can be catchy, but it is always memorable (especially to musicians). Styles are often imbued by a particular style of music, but a component of the person's personality can also come through to varying degrees. Then it depends on the quality of the personality – some are shallow, some are deep. It's like the sparkle in someone's eyes or the skip in their step. It's nice when you can hear this in their music.

Aretha Franklin is the classic example, and Luther Vandross is also a stylist. Some styles are old-fashioned and well known, some are contemporary, and some are a combination of old and new styles (for example, neo-soul music such as D'Angelo or Angie Stone). Occasionally you get the creative individual who can redefine a style. Rarely, you get the genius who creates a whole new style that becomes established in the consciousness of musicians everywhere. It is important to notice the extent to which the style sticks out on its own (which can be a good thing) or whether it fuses with the rest of the music.

Examples of style based on an instrument sound include Paul McCartney's style, which is partially based on his Hofner Bass sound. Earth, Wind & Fire's signature is their horn sound. An arrangement could even be the basis of the style – for example, the harmony vocals of Manhattan Transfer.

In addition to performance and instrumentation, a style is defined by a combination of each of the other ten aspects. When you examine where the style of a person or group shines through, you will always find it in one or more of the 13 aspects. James Brown is the classic example. You can hear his style in his performance, but it is also evident in his rhythms and arrangements.

Greatness

Greatness is the "goose-bump" factor. It is really a catchall phrase for any other important factor that people look for in music, and people hold a wide range of values. You should never let a performance go . . . until it turns you on. Common values include sincerity, heartfelt feelings, and emotions. Most likely, you are in this business because you know what you like. At the very least, don't let a performance pass that you don't like. If you make sure that every single moment of the performance is incredible, at least in your eyes, chances are that the overall performance of the song will be great, but you still need to check and make sure. Ideally, you want to get chills (to the bone) from every performance you work on.

> **Perfection has no limits!**
> **Once you obtain perfection, you advance to the next level,**
> **and you can see how much better things could be.**

Once you have developed highly refined values for what makes a great performance, you then must begin to learn how to coach musicians. The better you are at pulling great performances out of musicians, the better the quality of your projects will be. This skill is often learned through osmosis (by watching a seasoned engineer or producer critique a project), but can also be learned by trial and error and lots of practice.

Hiring Studio Musicians

Occasionally, it becomes necessary to hire a professional musician to replace someone in the studio. If you think telling someone to take lessons is hard, try telling someone that you are replacing their part with another musician. As you would probably expect, this could be seriously dangerous for your health. The best time to do this is when the project is almost finished.

We produced an artist who wrote a really nice song and wanted to play acoustic guitar. However, he wasn't a studio musician, and it is really difficult to play an acoustic guitar perfectly in the studio. It just so happened that we were working with a major studio musician from Emmylou Harris' band who was recording in another local studio. We told the songwriter that we should have this great studio musician come in and play his part. Of course, he wouldn't hear of it; he wanted to play his own part. Even though we told him how great this guy was, he said, "No, I'm the acoustic guitar player." So I told him that we would pay for the guy to come in and record, and if he didn't like it, we would leave the decision up to him as to whether we would use it. We brought in the studio musician, and the songwriter spent a couple of hours working with the guy, showing him the part. Then the guy played the part like an angel – it was light years better than the songwriter's playing. We didn't even have to say a word. Of course, we used the new part because the songwriter loved it. It was also good because the songwriter got to show the musician the part and work with him on it, so he felt like he was still involved. And as producers, we didn't have to pay for it (figuratively speaking) in the end.

Another approach is to talk to the musicians in the band about getting a different style of player. For example, David Gilmour brought in Lee Ritenour to play a guitar part in order to get a different style on one record. On one of Steely Dan's albums, they purportedly had about 40 different guitarists play the lead part, and then they chose the best one.

A great studio musician will often play something light years better than what you could ever imagine. In the past great studio musicians have *made* the group they are playing for (and often they don't get that much credit). These musicians are great examples: Joe Hunter – Piano/Keys, Richard "Pistol" Allen, Joe Messina, Ariel Jones, Benny "Papa Zita" Benjamin, and Eddie "Bongo" Brown. They helped create hits for the Beach Boys, the Rolling Stones, Elvis, and the Beatles.

Quite often a producer will work with a solo artist with no band and will need to hire musicians. When hiring a musician, keep in mind the following factors.

- **Musicianship.** Can he play, can he play, or can he play? Forgoing any budgetary restraints, this is the most important consideration.
- **Personality and attitude.** Is the musician easy to get along with? Is he or she open to taking direction? Is he or she afraid to give input?
- **Songwriter.** If a musician is also a songwriter, they normally have more to offer.
- **Chops.** Does the musician practice all the time? Many musicians don't have good homework skills, and they often don't practice a part until they get into the studio.
- **Availability.** Is the musician too booked?
- **Reliability.** Is the musician reliable and trustworthy?

Some people look at personality more than musicianship. However, if the musician can play, you can sometimes put up with certain personality issues. More than anything, the key consideration is, "Will the musician put the music first, no matter what, so it doesn't suffer?"

Homework

Do your homework while at a rehearsal and again at home while listening to the recordings of the songs.

1 Note each player's level of expertise. Look for any major weaknesses. Think about how you might deal with them.
2 Make sure the band is well rehearsed. Or, if you're recording MIDI (sequencing), all parts should be worked out prior to going into the recording studio. Do not wait to work out your ideas in the studio unless you are extremely rich! Studio time can be extremely costly.
3 Listen for any difficult sections where it is obvious that a musician is barely keeping up. Note these spots on your song map.

Approach in the Meeting

Critique the band's performance in the rehearsal. Make sure each person is playing at his or her capacity. If you see someone playing beyond his or her means, you can say, "Hey, I see that you're having a hard time with that lick. Be sure and practice like crazy so we don't have to spend eight hours trying to get it right when we're in the studio."

Musicians often get stage fright in the studio. Therefore, it is important that they practice as much as possible and get the part down before going into the studio. Or, you can simplify a part so it can be executed better and with more feel.

It is also a good idea to see the band play live. At a live show, you will normally see the entire range of a musician's talent. Then, when you get to the studio, you will know precisely how far you can push the musician.

In the pre-production meeting, sometimes it is necessary to address the expertise of the musicians. This can require some of your most delicate diplomacy. You might have a private discussion with the leader of the band first to get an overview of everyone's levels of musicianship. As you could probably guess, the big problem is when you have someone in the band who could really use music lessons. It's really hard to say to that person, "You could really use some lessons." Even if you end it by saying, "What do you think?" you might not get a positive response – to say the least

One solution is to ask the entire band to take lessons. Tell them that most musicians in major bands always take lessons (using coaches) and never stop learning! Even major drummers take lessons. Often major musicians learn by stealing (borrowing) ideas from other musicians. We once saw Phil Collins go ask Wynton Marsalis' drummer what drum lick he was playing.

When it comes to being a musician, no licks are sacred. Top-tier guitar players who could be instructors themselves also take lessons. And singers are always taking lessons. Even Celine Dion has a vocal coach. Tell the band that you would like everyone to go out and take lessons. Normally, the first lesson is free; after that, lessons normally range from $40 to $60 each. Give the band members copies of the recording to take to the instructor to see whether he or she comes up with any suggestions. Inevitably, the instructor will come up with better ways of playing the parts that the musician is already playing. For example, the instructor might show the drummer how to play the kick a little differently in one section of the song. A guitar instructor might say, "Try playing that part this way on the guitar. It's easier and it sounds better." And, you can sometimes invite or even hire a vocal coach to come to the session.

By getting everybody to take lessons,
the person who really needs lessons
gets the help he or she needs.

Of course, there are some people who will refuse to take lessons no matter how much you urge them.

Aspect 11: Mix

We won't go into the details of the mix too deeply because we covered it so extensively in *The Art of Mixing*. Please check out the book. Again, the more you know about every area of a production, the better.

Structure and Values

A *mix* is what you do with volume, panning, equalization, and effects. The key point to know is that the mix must fit the song like a shoe. It should fit the style of music, and it should fit the song. In fact, each one the 13 aspects is a key guide for how a mix should be constructed. Every component can help you decide what to do with volume, panning, EQ, and effects if you listen to what the aspects are telling you.

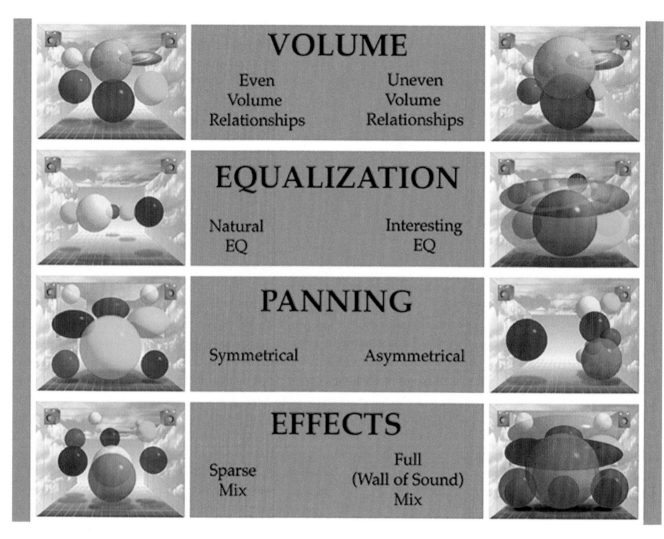

Visual 96
Dynamics Perspective

To start with, one of the key considerations is to decide how many effects the mix will have. Will the mix be very dry with barely any effects at all? (See Visual 97.)

Or, will it be full of effects, creating a wall of sound? (See Visual 98.)

**Visual 97
A Sparse Mix**

**Visual 98
A Full Mix**

Or, will it be somewhere between?

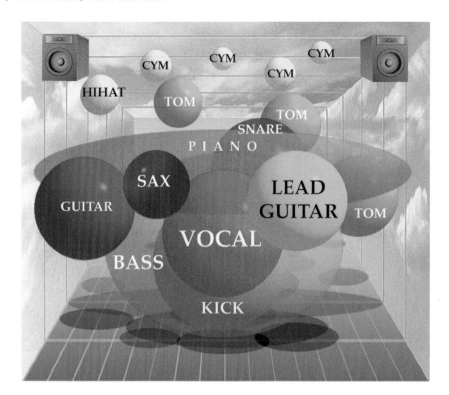

**Visual 99
A Mix Between Sparse and Full**

Remember, the mix may only be one small part creating a great overall recording; however, it is one of the most powerful aspects because the mix can be utilized to hide weaknesses in other areas.

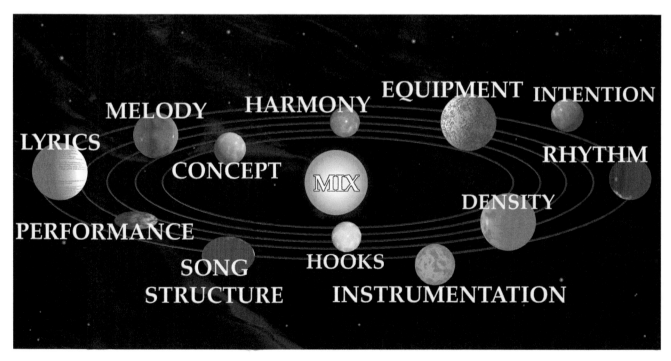

**Visual 100
The Solar System of the 13 Aspects**

Homework

Listen to each song and:

1   Design the overall mix that is appropriate for each song. Do you think it should be mostly dry, have lots of effects, or be in between? Figure out the type of reverb that you want on the drums and vocals. Decide whether you want flanging or other effects on any particular instrument sound. Write down why you think the style of mix you came up with is appropriate.

2   Pull out your lyric sheet and map out all of the places where you might want to have a delay on the end of a line, line, line. Note the exact words on the lyric sheet on which you want to put the delays. Use your song map to mark down any delays for any other instruments.

Approach in the Meeting

In the pre-production meeting, tell the band what you are thinking about creating for the mix. First, explain your ideas for the overall style of mix and see what they think. Perhaps the band will disagree, or maybe they will decide it is okay. The point is to figure it out so you can decide how the mix will be done. Then, your guitar player won't want to put really cool effects on the song that you previously agreed was going to be mixed dry. On the other hand, it's really cool when you're in the studio and the guitar player comes up with an ideal effect that totally fits the idea that everyone agreed on for the mix.

**The trick is to come up with a consensus for the mix in advance
so that people have a context within which to be creative.**

Once you have established the overall style of mix, present each of your ideas for specific effects on specific sounds. Tell the band what words you want to put delay on, on, on. and see what they think.

Aspect 12: Quality of the Equipment and Recording

As the producer, you make the final decision about where to record and mix. For this aspect, the quality of the equipment refers to recording equipment.

Structure and Values

The producer should make sure that all of the equipment is of the best quality possible and, even more importantly, that it is in good working order.

There are a few things to consider when choosing a studio:

*   **Analog versus digital.** As we'll discuss in Chapter 10, covering digital audio production, this is an important decision.
*   **Reputation.** Make sure the studio has got good recommendations.
*   **Price.** This is obvious – make sure the studio's rate meets your budget.
*   **Location.** It is nice when the studio is centrally located and easy for everyone involved to get to.

- **Sound.** Some studios are known for getting a particular sound. A studio often develops a reputation based on the groups that have previously recorded there. For example, Prince gets a particular sound at Paisley Park. Epic Records also was known for their sound.
- **Equipment selection.** It is important that you have the right equipment for the session. You should have at least four compressor/limiters; six is better (three stereo units). You should have at least three multi-effects processors and at least one high-quality mic preamp. You must also have at least one high-quality vocal microphone (a Neumann U87 is one example).

The quality of the recording is normally the responsibility of the engineer, but the producer should always oversee it. The more experience you have with studio procedures, the better. We highly (and unabashedly) recommend the book *The Art of Mixing* to learn the basics. It is important to make sure the engineer gets good levels on tape (not too low or too hot), uses good miking techniques, and gets no distortion or excessive noise.

Homework

Complete these steps before you go to the pre-production meeting. At home:

1 Make a list of all the high-end equipment you will need for the session.
2 Go to the studio where you will be recording and see whether they have all the equipment and instruments you will need. Go through your checklist and see whether you will need to rent extra equipment, such as microphones or mic preamps. This goes back to what we said about the band having a little bit of money.
3 Check out the size of the rooms and the number of isolation booths.
4 Check out the studio's amenities. Often these are nice features to keep musicians from getting too bored when they are waiting to play. Of course, this is not nearly as important as the sound, price, and equipment.
5 Make sure that all of the studio's equipment is in good working order. A lot of studios don't do maintenance until a major project comes in. Go to the owner and say, "I'll give you a $1,000 down payment from the band, and I want all the tracks on the multitrack working. And I want all of the channels on the board to work."
6 Pay attention to how the studio responds to your requests and concerns. This can be a good clue for how they might respond later if equipment breaks down. If they give you a hard time or they are not professional and respectful, go elsewhere.
7 Before booking the studio time, collect the schedules and availability of each band member. Then call the studio and schedule each session, and then confirm the schedules with each musician. It is often a good idea to check in with each musician the day before the session. Remember that we're dealing with *musicians* here! Leave as little room for errors and excuses as possible. Errors and excuses cost time and money.

Approach in the Meeting

If the band is involved in paying for the studio time, discuss the budget with them. Figure out how much you can afford for studio time and any extra equipment. Discuss the choice of studio with them.

Aspect 13: Hooks

A *hook* is a phrase that is catchy and memorable, especially to a large number of people. It strikes a chord on some deeper level that makes us remember it.

> Many people are now using the term "hook" to mean "the chorus" of the song. We are using the term "hook" to mean something that is catchy or meaningful and is memorable.
> Anything that sticks in someone's head after hearing a song.

**A hit song (or just a great song)
commonly has
multiple strong hooks on many different levels.**

Structure and Values

Just as we discussed the quality of the "concept" above, "hooks" can also be rated based on their appeal. Hooks can be rated on how memorable they are, or how deeply they touch you, or masses of people.

There is a wide range of hooks that can contribute to a successful production. In fact, each one of the other 12 aspects can contain a hook. Here are some examples of hooks in each aspect (although this is not an exhaustive survey by any means):

- **Concept.** A song in which every component is interwoven in the most beautiful way. Bach and Beethoven are the classic examples.
- **Intention.** An energy that comes across that brings you to tears, opens your heart, and blows your mind.
- **Melody.** A hummable melody that you can't get out of your head.
- **Rhythm.** A beat that stays in your head through the day.
- **Harmony.** A combination of notes and sounds that gives you serious chills. Occasionally, the harmony can touch a spiritual component in you.
- **Lyrics.** A line that is full of meaning and truth.
- **Song structure.** One that parallels an experience you have had. This might be a structure that shocks you into awareness or takes you to a new place that shows you a perspective you haven't seen before.
- **Density.** An arrangement that builds into a full monster or one that abruptly changes, waking you up again.
- **Instrumentation.** Sounds that are so unique that you say, "What is that? How did they make that?"
- **Performance.** A sincere heart-wrenching outpouring of the soul. (Peter Gabriel, Will Downing, and Patti LaBelle are just a few artists known for giving soulful performances.)
- **Mix.** A mix with a perfect combination of clarity and fullness or a mix that is manipulated in such a way that it becomes a solo instrument itself!
- **Equipment.** A pristine sound that makes your whole nervous system smile.

Creating Variations of a Hook

It is often nice to repeat a hook in various places throughout a song, but you have to be careful not to overuse a good thing. However, there are many ways to repackage a hook so that it comes through in different ways. Here are some examples.

- **Concept.** Use an antithesis or a parody of the concept elsewhere in a section of the song.
- **Intention.** Take your intention from the local perspective of the people in the room to a global perspective for all on the planet.
- **Melody.** Have the melody hook played with a lot of reverb by another instrument in the background (with a lot of reverb on it) in another section of the song.
- **Rhythm.** Have an unusual group of instruments play the rhythm hook elsewhere.
- **Harmony.** Have a synthesizer play the harmony of the background vocals in another section of the song.
- **Lyrics.** Take the hook and have it sung by the background vocals in a different way in another part of the song.
- **Song structure.** Have a section of the song that performs the whole song structure in a very short time.
- **Density.** Have the intro build up and break down in the same way as the whole song.
- **Instrumentation.** Put the same cool sound in the background elsewhere, but playing a different part.
- **Performance.** Have an unusually cool lick done the same way on different words later in the song.
- **Mix.** On the whole mix, take off the high and low ends with EQ and pan it all to one side, and then bring in another full-range, wide-panned mix of the same song on top of it. (This is called *shadowing*.)
- **Equipment.** Record one section of the song with low-fi equipment to highlight the good recording equipment throughout the rest of the song. This was done by Yes on "Owner of a Lonely Heart."

As you can see, the main idea here is to take the hook, change it a bit, and place it elsewhere in the song – commonly at a low level in the mix. When you have something really good, you shouldn't be afraid to show it off as much as possible.

Homework

Complete these steps before the pre-production meeting.

1   Define all of the hooks in the song. Determine which are the strongest and most important for the overall song.
2   Determine which one of the 12 aspects contains each of the hooks.
3   Is the hook at the beginning of the song? Do you want it to be there?
4   Think about how you might repeat hooks in the song or use variations of the hook.

Approach in the Meeting

Have an open discussion with the band about what the key hooks are. Point out any places where you might want to repeat or use a variation of the hook. Get the band's ideas, but also enlist them to come up with more ideas as a homework assignment.

Chapter 8: Higher-Level Concepts

In this chapter we focus on higher-level concepts of arranging and producing from both a creative and an industry-standard perspective. We look at how to create a cohesive style of music based on the song and its details. We also discuss how to order the songs on the album, and a bit about concept albums.

Creating a Style of Music

It is interesting to note all of the different responses you get from a band when you ask them what style of music they do. Often they will say it is a combination. Occasionally, they say that it can't be categorized. People are commonly fearful of pigeonholing their music into any one style. However, classifying a band's musical style can be important in order to define the music's market. Yes, it is nice when a song crosses over into other markets and becomes a hit in various genres. A classical pianist who plays punk might very appropriately call his or her style "classical punk." However, whether you like it or not, a song always has to "break" in one market. Therefore, it is good for a producer to help the band define their style of music for marketing purposes. After all, they are often concerned about the financial side also.

The point here is that if you don't do it yourself, someone else from the record company is going to do it for you, and it is much better to make your own decision. Also, if the record company starts to throw its weight around, be prepared to go to battle.

Structure and Values

Once you have defined the style, you can think about how you might want to adjust the band's approach. Of course, it is not good to try to change the band's style. You should always try to bring out the best of whatever they do best. This might mean something as subtle as changing the order of songs on the CD. The main point here is to get everybody on the same page.

Of the 13 aspects, rhythm, lyrics, instrumentation, and performance do the most to determine the style of music. For example, country and western is always in 4/4 time and you always know what the lyrics mean. For many styles of music, the instrumentation immediately lets you know the style. Just switch from radio station to radio station, and within two seconds you will know the style of music. However, on some songs it is difficult to determine the style until you hear the way the singer is singing. Other components that also contribute to the style include:

- **Concept.** You're not going to find a song about chainsaw murders in big band music.
- **Intention.** The overall energy that is held is a huge indicator of style, but not as specific as the concept.

- **Melody.** The simplicity or complexity says a lot.
- **Rhythm.** In most cases the beat alone can help you determine the style.
- **Harmony.** You often don't get three-part harmonies in rap.
- **Lyrics.** What the song says and how it says it – especially the language – can help you determine the style.
- **Song structure.** Certain styles rarely break out of the A/B song structure.
- **Density.** Some styles don't ever change the density (such as folk and punk).
- **Instrumentation.** This is the most telling aspect. You can commonly tell the style of music just by the sounds used.
- **Performance.** There's no twang in rap.
- **Mix.** Many styles stick to tradition. For example, relative volumes: New age, heavy metal, and alternative rock have more even volumes. Panning: Big band and classical have very strict rules. EQ: Acoustic jazz and country have very strict rules. Electronica pushes limits the most. Effects: Add a flanger to certain styles of music, and you'll go to jail.
- **Hooks.** The values as to what is good in something that is memorable goes a long way to defining the style.
- **Equipment.** Not much here that determines the style.

Often it is very effective to use influences from different styles of music to spice up your production. You can create a production that rides the fence of obscurity. For example, you might have song with various styles threaded through it. When someone who is into reggae listens to it, he or she might hear the reggae influence; a jazz person would hear the jazz; and a fan of Latin music might hear the Latin influence. In this respect you may be able to hit multiple markets at one time.

When you are dealing with a production, there are two different scenarios. One is when you must determine the style of music of a song that has already been written and decide the direction in which you would like to take the production. The second situation is when you are working to create a song and a production from scratch. You might have some lyrics and some rough music, but you decide the direction in which you want to take the music in the first place.

Homework

You should complete these steps before you go to the pre-production meeting.

1   Determine the primary style of the music. List any other influences. Think about the overall project and how it might best be marketed.
2   Go through each of the 13 aspects and write down what it is about each aspect that determines the style of music. Think about whether you would like to adjust that component to play down or highlight the style that is coming through the particular aspect. This might mean that you do something in the mix to highlight or hide an aspect. Take notes.

Approach in the Meeting

Explain to the band the importance of establishing a style of music. Be careful to point out that you are by no means trying to squeeze their music into any particular style. In fact, you only want to see what the reality of their music is so you'll know how it should be marketed. Point out that it is helpful to know how everyone would like to present his or her music on the market. It's really about creating a group identity upon which everyone can agree.

Point out some of the key components within the 13 aspects that contribute to defining the style of music. If you have any ideas about changing anything, present them and see what the band thinks.

Order of the Songs

The order can be approached from a few different perspectives. It is often thought of as a journey or landscape. The question is what is the emotional shape of the journey? How smooth or interesting are the turns? There are also many considerations based around your best and worst songs.

Structure and Values

First, try different orders and simply base the song order on whatever feels right. This is the most highly recommended approach. If you are doing a demo for a record company, be sure to order the songs by quality, starting from the best. Otherwise, use one of the best songs to start out the album, and try to make the *very* best song one of the first three tracks on the album. It's also nice to save a really good song for the end, in order to entice the listener to play the album over and over again. As we stated earlier, do whatever feels the best.

Another approach is to think of the album in terms of peaks and valleys. It is common to have two or three songs that gradually bring the listener up, followed by two or three songs that gradually bring him or her down. Some artists like to have multiple "up" songs and then occasionally have one slow to song to release tension. Occasionally, it makes sense to have all the fast songs first and the rest of the album slow. (This used to be more common when LPs were the deal.) A slow album will often get mellower or deeper as it progresses.

It is also nice if the songs are placed in an order that tells a story. One technique is to see whether you can make a story with only the titles of the songs. Occasionally, producers will consider the key of each song and how they progress.

Set the time between songs based on your feelings. Only use preset times between songs when you are in a serious time pinch. Listen and see whether the time feels good. Listen to the last ten or 15 seconds as the song fades out and note where it feels like the next song should start. Often it feels better to have a longer time after a song with a slower tempo. Sometimes you might not want any time between the songs, or you might want a cross fade. Check it out and set it however you like, because if you don't, the wind will set it for you.

Homework

You should complete this step before you go to the pre-production meeting.

Figure out the order of the songs, and then figure out an alternative order. Write down the reasons for your decisions. Often, you might decide to save this process until all of the songs have been mixed. That way, you can make demo CDs with various orders and pass them out to the band to listen to at their convenience.

Approach in the Meeting

Present the order of the songs to the band and explain your reasoning. Get the band's ideas and try to make a decision together. Make a demo CD for everyone to hear.

Concept Albums

A *concept album* is one on which every song is dedicated to an overall concept. *The Wall* by Pink Floyd is one of the most famous concept albums. The entire album is focused on the obstacles that are put in front of us over the years by society and life. Normally, a producer doesn't try to turn an album into a concept album unless it was conceived as such. However, if you are working with a band that is in the early stages of writing songs, you might try to detect whether the songs suggest an inclination toward a concept album.

A concept album is most effective when the overall concept is strong. This means that the overall concept is a strong hook because it touches people in a really deep way. If you are going to have an entire album devoted to one concept, it should not be fluff.

Chapter 9: Producing in the Studio

In the studio, the producer's primary role is to make sure that he or she gets quality creativity and energy recorded and mixed into the project every step of the way. Not only should every component on its own make you ecstatic, but it should also fit into the overall production.

Many producers continue the pre-production process in the studio. This is fine because ideas often don't emerge until you're in the heat of the session and the parts actually start coming together. However, the more you can get all of your pre-production done before going into the studio, the better.

It is much more cost efficient to do your pre-production in a place
where you aren't paying studio rates.

In the pre-production meeting, you came up with an overall picture of how you envision the final product. In the studio, it is a matter of keeping sight of your vision and still allowing the creative freedom to flow. When you let the creative process develop power on its own, you sometimes end up with something much more incredible than you could have ever imagined. The problem is that sometimes you don't end up with something incredible. Therefore, if you allow the session to go unguided, you run the very real (and common) risk of running out of money, time, and energy before you reach your goal of creating a quality finished product. (Also, bands often want to record as many tracks as the tape player or software will hold.)

The key is to find the general flow of the current performance,
sync up with it
and calculate what you need to do
to turn where the band is at
into what you are envisioning.

The acronym we use is CONCREF – controlled creative freedom – where you set certain limits or parameters within which the band can do whatever they want. The most basic constraints are time and money.

Critiquing a Performance

As previously mentioned, it is critical to treat your musicians with the utmost respect, no matter how bad they are (rather inexperienced). Otherwise, no one will listen to you.

Critiquing the Whole Band at Once

When you are recording multiple instruments at the same time, you need to be able to critique the whole band at once. However, this is a fairly advanced skill that takes years to master (although some people are born with it). For most people it is often difficult to hear whether every single note is right on for each instrument; often there is too much going on. Therefore, it is always important to ask each person in the band what he or she thinks of his or her own performance. You can normally count on the musicians to let you know about most errors. However, you must still listen to the playback very closely for any mistakes or areas that are less than incredible.

When you are listening to the entire band at once, one of your primary concerns is the consistency of the band's overall timing – it shouldn't speed up or slow down too much. This is your number one concern at this point. If you hear what you think is a problem, immediately write down the time on the counter so that when you play back the recording, you can go right to it and check it out. This can save you a huge amount of time in the studio.

Critiquing Individual Performances

Let's use a vocalist as an example, although these techniques also apply to other instruments. Imagine that the band has already played their basic parts and now you are going to record vocals.

First, make sure everyone has copies of the lyrics (especially you, the engineer, and the singer). Then, have the vocalist run through the song as many times as needed to warm up his or her voice. Be careful, though. Some singers give their best performances on the first take, and it's all downhill from then. You want to be especially aware of each artist's creative curve. The most common pattern is a singer who takes approximately 15 minutes to warm up, then gradually gets better for about one to three hours, and then gets tired and starts losing his or her edge. However, some professional studio musicians will actually get better and better for five to eight hours! It is really quite amazing. But then they crash completely. Some singers can only handle half an hour to an hour at time because their voices give out. After you work with someone for a long time, you get to know his or her creative curve, and you can get very good at telling when he or she has reached the point of diminishing returns.

> **Being aware of an artist's creative curve helps you calculate**
> **how much you can push him or her to get a great performance.**
> **At a certain point, he or she will simply get tired and/or irritated.**

After the singer is warmed up, have him or her sing through the whole song until you get a take that is approximately 70 percent happening. Then have the singer come into the control room and listen to his or her performance. At this point, you will get one of four responses:

- The artist hears himself and is shocked by how bad he sounds, so he goes back out into the studio and sings it much better.
- The artist hears himself and thinks it is pretty good, but maybe that it could be better in a few places.
- The artist listens and is extremely pleased with what he hears.
- The artist doesn't have clue or simply loses perspective when in the studio.

In each instance, you should have already checked out the performance yourself and decided how good it is. Calculate how closely you think the singer's perception corresponds to the reality of his or her performance. The problem to watch out for is when the singer thinks he or she was really great, but that wasn't the case. You then know that you may have to work with the singer more to achieve a great result because now you have to help him or her see how much better the performance could be. On the other hand, if the singer perceives that he or she really sucked when the performance really wasn't that bad, then you can breathe a sigh of relief because you know you won't have to work as hard. You still must make the final decision on the quality of every line, but now you are working together.

In fact, after the first song, you might not even bring the singer into the control room after he or she has completed the first take. It simply takes too much time. Only invite the singer into the control room when he or she has completed an entire song (or a section of the song).

There are two completely different approaches that you can take at this point. The first approach is quite common when recording a rap artist or any lyrics that have a lot of words or are sung (or rapped) really fast. These types of lyrics make it really difficult to punch-in (where you hit record on a single line or part) and preserve the cohesiveness of the performance. In this case it is common to simply have the singer perform the entire song multiple times – normally 3 to 5 times. In many programs the software has the ability to record multiple takes on one track (in Pro Tools it is called *playlists*). Then, you make a compilation track. Listen to the first line on each of the tracks and figure out which is the best.

COMPILATION TRACK GRID									
	Line 1	Line 2	Line 3	Line 4	Line 5	Line 6	Line 7	Line 8	Line 9	Line 10
Track 1				✓	✓					✓
Track 2	✓									
Track 3			✓				✓		✓	
Track 4		✓						✓		
Track 5						✓				

Chart 10 Compilation Track Grid

Then copy that first line from the best track over to a new track.

Continue going through each line and decide which track contains the best take. Copy all of the best takes for each line to a whole new track. In Pro Tools, it is a simple as going to the take that you like and copy the line you like. Then go back to the best take and paste the line in. When you are done you have a track with the best of all the tracks for each line!

However, the problem is that the compilation track may not sound cohesive because it is from different takes. We learned a technique from producer, Sonny Limbo, that takes this procedure to another level of quality. Have the singer listen to the compilation track.

Then, have the singer go out into the studio and sing a whole new take on a new track. Tell the singer, "Now go out there and beat what you just heard on the compilation track." You will almost always get a take that is light years better than any of the other original takes.

Make a rough mix of the final take, and make a CD for the singer to take home. Tell the singer, "Take this home, listen to it, and practice. Next week, come back and we'll do the real vocal." Inevitably, when he or she comes back the next week, you have a whole new singer!

The second type of approach is to fix one take, section by section. This approach is commonly used when you have a song with some time to punch-in between each line of the lyrics.

Again, you get a take that is at least 70 percent happening, and you listen back in the control room. While listening to the playback in the control room, you would think that if you wanted to be as efficient as possible, you would write down the counter time (on the multi-track) on your lyric sheet every time the artist makes an error – then you could go back into the studio and efficiently fix the problem spots. But no one ever does that. It is nice to be able to sit back and listen to the overall performance without focusing on precise words quite yet. The problem is that after listening to the playback, the singer goes back into the studio and says, "Let's listen again." Then you listen and you still don't know precisely what lines you need to fix in the song because it is so hard for everyone to remember all the lines that need to be fixed and exactly what was wrong with them. This can be a huge waste of valuable time.

Instead, what you want to do is follow a specific procedure that will save you a huge amount of time. Have the singer go back out into the studio and put on the headphones, and then tell him or her that you are going to go through the song and fix the vocals section by section. Even if the singer says, "I want to sing the whole song again," don't allow it. You have already agreed that the song is 70 percent happening. Tell the singer you don't want to lose all the good work that he or she has done already. "Let's just fix the parts that are screwed up," you might say, "and keep the parts that are fine." Sometimes, it is necessary to force (or convince) the singer not to record the whole thing again.

Occasionally, you will get a singer who does not like to do punch-ins. (This is the case with Aretha Franklin and many jazz musicians.) These singers will only sing through the whole song. In this case, try to get a performance that is at least 90 percent happening. This also applies when you are recording straight to two tracks and you are mixing as you are recording. In this case, the vocals are not on a separate track, so you cannot punch-in anyway.

Again, have the singer go back into the studio and put on the headphones, and then play only the first verse or section of the song. When you start playing the first verse, you can go to work more intensely than at any other time in the studio, including mixing. This is when you have to concentrate intensely. This is *the* most important part of being a producer.

When you play back the first verse, you will listen for four main things (and the fourth thing has many parts to it). The following sections detail these four concerns.

How Many Lines Are There?

The first thing you listen for is how many lines the verse or section has. Most verses have four lines, but sometimes they have eight and occasionally they have six. Rarely will verses have an odd number of lines. Therefore, it is normally very easy to count the number of lines.

Sometimes someone will think of a verse as four lines when you think it is eight. It doesn't really matter how many lines there are, as long as everyone agrees. Normally, it is better to divide the lines into smaller sections (eight, for example) so you don't end up talking about the first and second half of each line. It doesn't matter if a line is not a complete sentence or thought.

In the digital audio workstation, you can look at the waveform and easily see exactly how many lines there are.

How Long Are the Lines?

The next thing you want to listen for is how long the lines are. You don't need to know the precise minutes and seconds, just the relative length compared to the other lines. Most of the time, all of the lines are the same length. The second line might be the same as the first line. Or, it might be double the length of the first line, or perhaps just a little bit longer. Maybe one line is really short, with just a couple of words. Again, the lines normally are all the same length. If they are different, just take note of it. You simply need to know the relative length of each line.

Again, when working in a DAW, it is obvious when you look at the relative waveforms.

How Big Are the Spaces between Lines?

Listen to see how long the space is between each line. You do this so you know whether it is easy to punch-in. There are three lengths of spaces. The first is so long that anybody could punch-in without knowing anything about the song. With this size of space, if you want to punch-in on line two . . . let's go. In a DAW it is easy to just select where you want to record by dragging across the section. In this case you can start select so there is a bit of time before and after the line being fixed . . . in case they come in early or go longer during the new take.

The second length of space is extremely close. It is tight, but you can still do it, as long as the singer does it almost the exact same way and doesn't come in too early or go too long (into the next line).

The third length is no space at all. This can be really tricky to get the musician to sing or play exactly the same length as before. This means if you have a bass line that is playing nonstop throughout the entire song, whenever you punch-in you will have to record to the very end. When this is the case, tell the musician to stop playing if he knows he has made a mistake. You can then punch-in and continue to the end (or the next mistake).

Having noted the number, length, and spaces between the lines, you know where you can punch-in and where you cannot. If the band says they want to fix something in the second line of the second verse, and you know there is enough space, you can then punch-in automatically without having to listen to the punch-in point over and over. And then the band goes, "Wow! This producer is happening!"

How Good is the Performance?

The fourth thing you are listening for is the quality of the performance. As discussed in Chapter 6 in the "Aspect 10: Performance" section, you are critiquing timing, pitch, dynamics, technique, style, and greatness (the goose-bump factor). This is a lot to listen for. Not only do you listen for the number and length of each line and the size of the space between the lines, you also check out every line to see whether the timing is happening, the pitch is on, there are too many or too few dynamics, the technique is right on, and whether it gives you serious chills. Ultimately, there is one important question to always ask yourself on every line:

Is the performance good enough to keep?
Is it great
or are you going to redo it?

The procedure works this way. Listen through the first verse and critique the performance. First, critique the overall performance. Perhaps the singer needs some coaching on his breathing techniques. Perhaps he needs to project more. Then, critique each line and note whether you can easily punch-in to it. If you have a comment, note the precise line. Ultimately you want to be able to tell the band what word has the problem – to the point of whether the problem is with the beginning, middle, or end of the word. Push the talkback button and discuss what you both think about the lines in the section you just heard.

When you are making comments, always be careful to give only constructive criticism. It does no good to tell someone that they need lessons, or should give up singing. Regardless of how bad you think a musician is, it is your job to help bring up to the next level of expertise. The more you can help the artist to excel, not only will you end up with a better product, but you are helping them to grow. Even if you don't see this an altruistic value, they will almost always appreciate your help, and will be more likely to return to you on their next project, and will be more likely to recommend you to others. Always remember that your goal is to try to inspire people. Be careful with criticism. Always edit before you speak.

For example, the singer might say that the first line was really pretty good. And maybe you agree that the first line *was* really good, but it should be incredible. It is the first line of the song; it should be perfect. And then the singer says that the second line is a little bit weak, but you reply that there is some serious emotion in that line. It actually shows a lot of vulnerability. And the singer says that he doesn't know what to think about the third line, so you point out that the third line actually had a pitch problem. You tell the singer to listen back the next time to see if the pitch problem was really that bad. Then the singer says the last line was the best that he has ever done. You point out that there was a small timing problem with one word but, "I know what you mean. It had an essence that was amazing. Let's check it out and see if that timing problem really bothers me."

If you were to map this out, it might look something like this:

Line	Singer's Perception	Your Perception
Line 1	Really pretty good	Really good, but should be incredible
Line 2	Weak	Shows emotion and vulnerability
Line 3	Don't know	Has a pitch problem
Line 4	Best ever	Essence amazing, but minor timing problem

After you both have pointed out your perceptions, you then listen to see whether what the singer heard might be right and whether your perception still holds true. Often when you hear something a second time, it doesn't even bug you. On the other hand, sometimes a small error will bug you more and more every time you listen back. The error seems to grow in size and irritation.

There are two types of problems in this world.
The first problem is the type like a hole in your sock.
When you first put on your shoes, you go, "It's a little cold down there,"
but in an hour you've completely forgotten about it.
The second type of problem is like a pebble in your shoe.
It doesn't go away. In fact, if you don't get it out, it causes a blister!
It gets worse and worse until you get that sucker out of there.

Have the singer listen for what you just heard, as you listen for what the singer just heard. Play the song and listen to see whether that first line really is good enough. See whether the second line really has that much emotion. (The singer is listening for it, too.) See whether the third line has a bad pitch problem, and whether the timing problem in the fourth line is that bad or if the emotion of the line carries it.

It is often nice to be able to see the singer's face so you can see how he is reacting to each line as he sings it. You can not only see when they like what they did (or not), but you can also see whether they are physically in the groove.

Then, negotiate whether each line is good enough to keep or whether it should be redone. Make a decision and fix the lines that you both agree need fixing.

Focus and Concentration

It takes a huge amount of concentration and discipline when you really get into critiquing a song in detail. It's not an easy thing to do. Imagine focusing this intently on fine details for eight hours. You sleep really well at night because you're exhausted! Getting really good at this level of concentration requires some serious practice. The cool thing is that the more you do it, the better you get at it.

To listen intently at every single moment is a lot of work, which is why professional engineers and producers don't work for free. It is serious work. You don't go into the studio to have fun (although that's not against the law). When you're in the studio, it's concentration time. You don't want to put out a CD with both BS *and* your name on it. Don't let any BS get by you. Don't fool around. Why spend your time if you're not going to do your best?

It is interesting to observe the different amounts of time that it takes for people to reach this high level of concentration and focus, as well as the different levels of concentration that people reach. We have observed that in a room full of people, some get into it deeper and quicker than others. People with more critical listening experience dive right into a very deep level of detail. It takes others a bit longer to get their minds around such fine details.

We have found that on some days it takes longer to get into focus than on other days. On a good day, it may take only a few seconds to a minute to get in. It often depends on how tired you are and what your overall state of mind is. When you are stressed, it might take you as long as 15 minutes to really get into a song. However, we have discovered that you can be emotionally distraught and feel totally out of it, but you still are able to immediately sink into deep concentration. Therefore, it seems that the mental capacity required for such deep concentration is not *necessarily* connected to your emotional state.

As a producer it is important to get to the point where you can immediately hear every fine detail, and pay attention to the overall essence at the same time. You can also learn how to guide others to this level of awareness (discussed in detail in Chapter 13, "Moment-to-Moment Awareness").

The most interesting phenomenon is that you can actually raise the level of someone's concentration simply by osmosis. It is a known fact that a large percentage of people in a room will hear the same detail as the most precise listener in the room – especially if that person points out what to listen for. It is a bizarre phenomenon. We've encountered it many times ourselves, when we were in the room with major musicians, engineers, or producers. It's as if you start hearing what they are hearing. If you are paying attention, your ears will often open up to the same level as the finest listener's. This explains why it is so difficult to develop a good ear on your own. Spend even a few weeks with a "major ear," and you will hear things you have never heard before.

As a producer, it is important to include everyone in the room in this critiquing process, especially the recording engineer. It is nice when both the engineer and the producer are totally rested and up to this high level of concentration. But when you can get the band concentrating at this level as well, the level of focused energy in the room can do wonders for a performance. Get a room full of people listening intently, and it can be scary.

Now you, as the producer, need to keep track of more than only the comments of the singer and themselves. As you get everyone involved, you need to keep track of all of the comments that everyone makes for each line of the song. Listen for what each person comes up with and see whether the same things bug you. Get everyone in the room to listen for each point that anyone brings up for each line in the song. When you get everyone involved in fine details like this, the project ends up producing itself. You still want to guide the whole process and retain control, but the group usually will reach a very high level of consensus.

**When everybody is critiquing everything so intensely,
musicians can't help but get a huge contact high
from all the energy being put into the process.**

With all of this energy, there is no way you could come up with a bad performance unless the band is having a bad day. You put this much energy into it, and they are going to get it. You will get the best you can get out of them.

**The job of a producer is to get everybody
up to this level of concentration and focus.
This is what producing is about
when working in the studio.**

The key is to create an atmosphere of intensity in the room that is focused entirely on every detail of the music. This isn't to say that you can't have a good time. Humor is important for letting off steam and tension.

The first step is for *you* to listen intently. Often this will entrain others into hearing what you are hearing. The second step is to get everyone in the room to help you listen. Do not allow any distractions. If you hear anyone talking or not paying attention, turn around and say, "Hey guys, help me listen. Do you hear a pitch problem? I could really use some help here." It is helpful to cue everybody in the room about precisely what you are listening for. If necessary, give everybody the list of six things to critique in a performance (timing, pitch, dynamics, technique, style, and goose-bumps). Let everybody know that the key point you're looking for is whether each line is good enough to keep.

You need to make it clear to everyone in the studio that you are there to work. Tell others in the studio that if they need to talk, please go upstairs or wait for a break. After you establish this atmosphere, it will become easier to concentrate as everyone in the room joins in.

When everybody's into it, the vibe in the room is intense! Everybody is focused and listening closely. That's what should be happening in the room at all times.

The most incredible thing is when you get a room full of major musicians or players who have been doing this all their lives – the vibe in the room is unbelievable. The intensity of the concentration is magically powerful. It's an amazing thing. Again, everybody in the room essentially gets pulled up to a really high level of awareness. When you have a whole room full of professional musicians, there is such a high frequency of intensity that *it is nearly impossible to get a bad take*.

That's what it is all about – getting everybody into a space of detailed concentration and focused attention. *That*'s producing in the control room.

Other Contributing Factors

A number of factors contribute to the decision of how much time you will spend trying to get a great performance. It is the producer's responsibility to gauge the amount of time spent on refining a particular performance, in order to bring a project in under budget. Regardless of the circumstances, everyone wants a basic level of good quality. However, after obtaining this basic level of quality, there is only so much you can do to get an excellent performance. It depends on several factors, which are detailed in the following sections.

Budget

If the band can't afford the time to perfect a performance, and the record company is out of money, there is nothing you can do unless you, as the producer, want to invest your time and money – or you own the studio and are extremely generous. You just might have to move the session along and spend less time in the studio.

Deadlines

A deadline, such as a meeting with a record company, a mastering appointment, or even Christmas (especially when a project has to be ready for holiday sales), is one of the primary destroyers of project quality. However, sometimes these deadlines cannot be avoided. If a group has a limited amount of time, you might have to accept a performance that is less than perfect.

Purpose of a Project

Obviously, if a project is destined to be a CD, much more refinement is in order. Vinyl is final, and every album is a part of your reputation. If the project is being done as a demo, then you might let a less-than-perfect performance pass as acceptable.

Expertise of Musicians

The quality of musicianship makes a big difference in the amount of time it takes to get an acceptable performance. You would think that the worse the players are, the longer it would take, but this is often not the case. Many times great musicians take even longer because they know how good a performance can be.

Apparent Musical Values

Different people hold different values for their music. For example, a punk band might focus on energy instead of perfect pitch. An R&B band might care about the spatiality of the sound. A rap group might be concerned mostly about the "boom." A jazz combo might emphasize the interaction between the players. Often these values will determine whether a performance is acceptable. It is frequently fruitless to spend too much time on an aspect about which the band could care less. However, you should demand certain basic levels of expertise.

Determination

The amount of determination that a band brings to a project affects the time spent working on a part and the quality of the final project. Often band members don't realize how much work it takes to get a performance perfect or great. Musicians can easily get frustrated or fatigued to the point where they say, "Good enough, already!" You should always try to inspire everyone to work harder and longer until the performance is as good as possible, but you can only push musicians so far before they become irritable. It might help to simply point out that it is normal to take a long time to get things right and that professional musicians often take days to perfect a performance. This can help inspire people to push themselves to do their best.

On the other hand, some musicians are so determined to get a perfect performance that they never stop. In the beginning these people might drive you nuts, but you'll soon realize that with this kind of perfectionism, you will end up with an incredible performance. Subsequently, when people listen to the project they will say, "Wow, you produced that?" Therefore, you come to appreciate the obsessive musicians.

Chapter 10: Overall Production Goals

A great production commonly has multiple hooks that all work together. In 95 percent of all hits, the 13 aspects support each other. The concept, melody, rhythm, harmony, lyrics, song structure, density, instrumentation, performance, mix, and equipment each support each of the other ten aspects.

In 95 percent of all hits,

every aspect

supports

every other aspect.

Your concept supports each of the other 12 aspects.

Your intention supports each of the other 12 aspects.

Your melody supports each of the other 12 aspects.

Your rhythm supports each of the other 12 aspects.

Your harmony supports each of the other 12 aspects.

Your lyrics support each of the other 12 aspects.

Your song structure supports each of the other 12 aspects.

Your density supports each of the other 12 aspects.

Your instrumentation supports each of the other 12 aspects.

Your performance supports each of the other 12 aspects.

Your mix supports each of the other 12 aspects.

Your hooks support each of the other 12 aspects.

Your equipment supports each of the other 12 aspects.

Everything supports everything else.

(Sorry to waste so much paper driving this point home, but you get the idea – this is an important point.)

Combining the 13 Aspects to Create a Major Production

Each component must shine on its own, but it must also fit with the others. When everything fits, a synergy results that absolutely creates more than the sum of the parts. This is called magic. Every aspect of a song has to support the theme or story, unless the song is intended to be fragmented and abstract for a specific purpose or effect. An excellent illustration of this can be found in the documentary, "Standing in the Shadows of Motown."

Putting all the components together creates an overall production that you and a wide range of others will perceive as cool. The truth is that a great production lies more in the relationships between the 13 aspects than it does on any one of the components. The magic lies within the relationships of the aspects, and there are many combinations. The key is to create as much magic in each component as possible, and then see what new magic you create inadvertently by combining the aspects. Quite often a synergy will emerge that is much greater than the sum of the parts. The key is to always be on the lookout for any hidden magic that might emerge. Then, think about using that magic as the main hook and redeveloping your song around this synergy that has now obviously become the central, shining highlight.

Make sure you are flexible enough
to redirect the production
to follow
what the song has come up with.

In a similar vein, it also is important to be able to throw out tracks – even tracks that you have spent a good amount of time working on. Perhaps after repeated listening you realize that you hate a guitar part that you spent two days recording. As a great producer, you must learn to let go. Get rid of it.

Likewise, the hardest thing to do is to actually throw out a whole song and start over. It can be difficult to admit defeat, but it can also mean that your overall project will then be a success.

Same thing when it comes to a less-than-perfect mix. Don't hesitate to spend the time to redo the mix. Don't be attached to your work.

Overproducing

There are many ways to overproduce a song, but the result is always the same – the essence of the song is lost.

One of the primary ways to overproduce a project is simply to add too much stuff. Less is often best. Listen to how sparse the arrangement is on many hit songs. Often the trick is simply to take the one key component that makes the song great and put it right out front by itself. The key is to always ask yourself, "What is the essence of this song?" Make sure you never cloud it over.

The most common problem is when a musician is worked to death and delivers a performance that is technically perfect but has lost the sincere feel and emotion it originally had.

Too much elaboration is also a culprit. Any time you come up with a logical idea to change something, always ask yourself how it will affect the natural feel and emotion that is already present.

Subtle essences commonly are lost
in big ideas
that are actually quite smart!

Another way to overproduce a song is to add effects that don't fit. Simply keep in mind the essence of what the song is about. What is the primary message that the song is trying to get across. If you keep this in mind and always ask yourself, "No matter how cool an idea might be, does it fit? Does it add or detract from the song's message?"

High-Level Producer Values

Many major producers have developed and follow high-level values that cut across all styles and genres of music. One of the most common values is that not a single component of a song should lack anything in the producer's eyes. Others may disagree, but you can't please everybody all of the time. This means that you get as good a quality as possible by any means necessary, even if it means hiring different players.

The music should never suffer.
—Duke Ellington

It's the song that dictates the mix.
—Bradley Ditto

Look for these high-level values in every song you have time to listen to for the rest of your life. When you find such magic, put it into words to help you remember what they did. Then, figure out how they created the magic. Ask yourself, in which of the 13 aspects is this magic? Then you now have that technique in your bag of tricks to use on a future session. The top producers in the world are often professional thieves (in the best sense of the word).

When you get down to it there are two main overall considerations that greatness is based on:

— The history of how the style of music has been produced and mixed.
— This is a conglomeration of the overall consciousness and energy that has been popular for many years.
— The deep feelings of the creators of the music.

In the beginning people did not create music to please others. They created to please themselves. Or, more importantly, they had no thought of pleasing anyone. The music just came through. All of the values that developed for the history of any genre has always started with the values and inspiration of a single person (perhaps connected to source).

To learn the high-level values of music creation and producing
you can study everything that is out there and find the magic.

Or, you can simply your trust yourself and
pull from your inner core.

This is where the true magic is!

PART 6

Creativity

Chapter 11: Tapping into the Source: Inspiring Creativity

As a producer you will often run into a situation where the creative juices have dried up. This could happen with the band, but it can also happen with yourself. This chapter provides techniques for getting inspiring creativity. The cool thing is that once you learn how to tap into the source, you get to the point where you can access it at any time.

We have found that there are two general types of creativity – creative problem solving and artistic creativity. Creative problem solving often comes with a level of tension or stress, as there is often a deadline looming. Artistic creativity does sometimes also have a deadline (especially for those who are producing sound for video for film), however normally there is more time to allow ideas to develop.

Creative Problem Solving

Creative problem solving comes into play when you have an instrument with a hard-to-find buzz, a scheduling snafu, or any type of miscommunication issue. In this case, the creativity required to solve the problem must arise in the face of stress and adversity.

Problem-solving capabilities generally require getting as many of the facts together as possible. Then, you must define precisely what you are trying to achieve. You must also define the primary obstacle blocking the solution and try to come up with alternative routes to achieve a resolution.

The most important first step is to realize that everybody is ultimately on the same side trying to figure out a solution. Different people may have different perspectives, but everyone is working towards making the project as good as possible. This can help take the "personal" out of it.

It is helpful to simply roll the situation around in your mind awhile and look at it from different perspectives. In one respect, the solution is actually engineered based on the relevant information. Often it is a matter of finding the right tool.

One of the keys to problem solving is to stay calm under pressure. Sometimes it is just a matter of breathing. Sometimes it is a matter of taking a break.

Artistic Creativity

The second type of creativity could be called *artistic creativity*. Psychiatrists and philosophers commonly indicate that this source of creativity is due to some sort of angst (anxiety) or pent-up energy that needs to be released. The creative act serves as a catharsis of sorts, releasing the energy and freeing the person. It is generally agreed that when we view or get involved in a person's art, it has the potential to provide us with the same catharsis that the artist experienced in its creation.

You can imagine a guitar player saying, "I just play, man." However, it does beg the question, "What makes some people more creative than others?" But, more importantly, why are some people capable of great works of art and others are not? What is it that makes a person capable of great art?

First, artistic creativity is similar to creative problem solving in that you have a set of circumstances, and you often use a tool in the creative process. However, the solution in this case is the art itself.

Many books explore the creative process and what it takes to enhance it. We'll discuss some of these in a moment, but the most important question is, "What is creativity in the first place?"

Some say, "It's all been done before." We are only rearranging all that has been done before. It's all sampling! On a subconscious level, you could say that everything has already been created, and we are only remembering it. There are only so many notes that can be put together in so many ways in music. There are even fewer ways that the four tools of volume, panning, EQ, and effects can be put together to create a mix. Creativity is really just a way of reorganizing old information.

But creativity is more than that. There is some lubricant or other energy that often participates in the process of this reorganization. Some see the lubricant as Source Energy. Sometimes it is ascribed to the other side. Some blame angels or aliens. Some simply call it divine intervention. Others don't try to ascribe it to anything; they simply look at the purity of the essence coming through.

It's all about
the level of
purity of the essence
coming through.

This is the type of vision that a great producer must develop and refine – the ability to see multiple levels of essence. As we discussed in *The Art of Mixing*, what it really comes down to is being able to categorize different levels of beauty or intensity on a purely objective level.

The key is to get really good at seeing the magic. When it comes down to it, it's really about knowing what magic is in the first place, and knowing what it feels like.

We use the term "magic" to refer to whatever it is about music, songs, and the use of the equipment that turns you on. This could include beauty, intensity, massiveness, chaos, sincerity, love, or just *the flow* – whatever gets you going!

The trick is to develop a radar for finding magic, but it is a little more detailed than that. You must become totally immersed in this world of magic. You must become a master of the subtleties of magic. You must develop a highly refined hierarchy of magic. You need to categorize the magic you feel in your heart and entire being.

What is the magic in music? Think about it. Begin to develop a list of some of the examples of magic in music. In fact, you probably already have a pretty long list, but you might not have focused on it yet.

Here are some examples:

- A rhythm that moves you, or makes you move.
- A melody that touches your heart for some reason.
- A lyric that seems to fit exactly what is going on in your life, especially one that gives you a new perspective on a situation or helps you get out of a rut or vicious circle of thought.
- Parts of a song that seem to fit together in a way that creates something greater than the sum of the parts.
- A song that just seems to fit your mood perfectly.
- A singer who is totally putting his or her heart and soul into singing a song.
- A perfect combination of simplicity and complexity that doesn't bore you but also isn't too much.
- Harmonies that are just amazing.
- A performance that obviously took a huge amount of experience, practice, and talent to pull off.

As a producer, you become more and more familiar with all of the magic that music can create. It is a lifetime process. Then you get good at organizing magic at different levels of perception.

Pay very close attention to the fine details of what turns you on in a particular piece of music. Some things affect you one way; others affect you another way. Some are extremely deep; some are just cool. Some are fun or titillating; some touch your heart. Some are short in duration; some grow on you over a lifetime. Some touch you and your friends; some touch every soul in the world! There are various levels of feeling. It's like seeing the magic in a simple blade of grass, and then seeing the magic in a green field or a range of green mountains. They all have their beauties, but the intensity of their beauties is different. Once you have different levels of magic clear in your mind, you can make a better judgment about what to highlight in a song or piece of music.

Of course, magic is different for different people. You should not only learn what turns *you* on, you should also look for what turns others on. For example, whenever anyone says that he or she *really* likes a song, pay close attention and try to see what it is that lights that person up. Look for the magic that other people see.

Really, when it gets down to it,
what you are doing as a producer
is simply
balancing different levels of magic.
It's like painting with magic.

The magic in your production are like the highlights in your painting. Not only do you want a nice balance of highlights, you also want them in just the right places.

When you find magic, name it! Naming it helps you corral the idea and get it into your memory bank so you can use it for yourself later. Add it to your producer's bag of tricks.

The cool thing about this is that once you develop a radar for magic in music, it starts spilling over into the rest of your life. You start to develop an acute sensitivity for aesthetics in all that you see and do. Nature becomes more beautiful, any art becomes more intriguing, and people become works of art to admire. Every great artist in the world develops a radar for magic. It is an acute aesthetic sensitivity.

It's called being an *artiste*.
This is really what makes a great producer.
Anybody can make music,
but only a few learn how to create true magic.

Once you become a master of the magic in music, the best way to use music to enhance the magic becomes obvious. You just feel what will work the best.

This is what makes a great producer. The process doesn't happen overnight, but the journey is a pleasure to take.

Techniques to Get the Creative Juices Flowing

Here are some exercises to help you access the well of creativity. These exercises fall into three main categories:

- Anything that puts you in touch with the silence within you.
- Anything that touches you.
- Anything that whacks you out or knocks you off center.

Quietude

The number one source
of creativity
is
silence.

Meditation is known to enhance creativity. Some say that creativity is simply letting the spirit come through you. If this is the case, silence allows the spirit to come through most clearly. This is not to say that it can't come through in the middle of noisy chaos, but silence often creates space for ideas to come through.

Sweetness and Feeling Good

Anything that works to make you "feel" something in a nice way or touches your soul will often bring you into place where the creativity begins to flow.

- **Nature.** Communing with nature and animals can inspire creativity – especially aspects of nature that are moving or flowing. I find that a river or the ocean is especially helpful.
- **Listening to music.** Often listening to music brings you into that space where creative ideas flow.
- **Sex, but especially love!** Almost any type of positive connection you make with another human being can get the creative juices going.
- **Exercise.** Releasing endorphins can put you in good frame of mind.
- **Just playing.** One definition of playing, from Webster's dictionary is "to move or function freely within prescribed limits." For kids, playing is their way of learning about the world.
- **Laughter.** Stretching the cheeks is known to release tension and allow creativity to flow.
- **Sleep.** Not only can a good night's sleep help, but often dreams are an incredible source of new ways of looking at things. And of course, don't forget daydreaming.

Stress and Chaos

Anything that is shocking or throws you off from your normal self can also trigger more creativity. Situations leave you in a place where you see things from a new and different perspective. Often the resolution of such conflicts can be the source of creativity inspiration.

- **Loss.** A breakup with your mate or the death of a loved one can bring out your creativity. Even the loss of physical objects (such as your house or even your keys) can unleash a gush of creativity.
- **Personal conflicts with others.** Any type of conflict can lead to creativity.
- **Drugs (including alcohol).** There are the rumors, at least. We are definitely not advocating them. In fact, it has been proven that any creativity that can be accessed through drugs can just as easily be accessed through other means without the bad side effects.
- **Lack of sleep or other sensory deprivation.** Often you don't feel like yourself and this can lead to thoughts that are out of the norm.
- **Fear.** Fear of moving into a relationship because your heart has been broken, fear of world destruction, fear not being good enough or loved . . . Almost every type of fear has been dealt with in songs.
- **Any trauma.** Trauma, such as an accident or a near-death experience, can be an incredible source of inspiration.

Now, whenever you seem to be stuck, come back to this chapter.

Chapter 12: Balancing Creativity and Mass Appeal (Industry Trends)

Now that we've explored the creative process, it almost feels sacrilegious to discuss creating music that appeals to the mass audience. But if you are to become a great producer, you have to at least be aware of what appeals to the masses. This is not to say that you have to create music or productions that cater to the mass audience, but you should be aware of it.

The Eccentric or Creative Genius versus Tapping into the Mass Audience

Most people think that subscribing to mass audience appeal means dumbing down the material. Sometimes this is the case, but just as often it is not. Plenty of really unique and creative bands are hits. Pink Floyd, Radiohead, Nine Inch Nails, Art of Noise, Snoop Dogg, Jewel, Tori Amos, and even Eminem are doing some very unique stuff. There are some artists who were considered cutting edge when they first came out, and have now become standards, such as Little Richard, Sly Stone, Elvis Presley, D-Train, Led Zeppelin, Prince, Madonna, and Tupac Shakur.

The boy bands are what most people consider an extreme case of catering to the mass market instead of letting creativity dictate the music. Many people think that using formulas to create music is evil. But how evil is it? True evil is really when people go out and kill each other. Hitler was true evil. Making any type of music (no matter how bad) is much less evil than making bombs. And, as many people say, aren't we just recycling all the ideas that are already out there anyway? It's all been done before.

On the other hand, there is something to be said for music that is totally inspired by an honest feeling or sense of spirit versus music that was conceived with making money in mind.

But what do you do? It *is* a business and either you play by their rules or you don't play at all. Of course, you can work to change the rules, and there are many major artists who have lobbied, harassed, and even sued record companies and have subsequently made the world a better place for all creative musicians. (Laurie Anderson, John Fogerty, Frank Zappa, Don Henley, and Peter Gabriel are just a few examples of such artists.)

Often the job of the producer is to balance the needs of the record company with the desires of the band. He or she often functions as a mediator between the executive producer and/or the record company and the recording artist. To be effective, the producer must help the band understand that the executive producer and the record company expect to see lucrative returns on their investments, while still allowing the artist creative freedom. This combination of business and freedom is called *controlled creative freedom* (CONCREF).

For instance, the producer cannot allow an artist to include a three-and-a-half-minute solo on a song destined for radio airplay. Standard airplay is no more than four minutes and 15 seconds, and the song needs to say something within that limited time. Likewise, you are not going to get a song that contains the "f" word or the "n" word on mainstream radio. The professional producer might have to constantly remind the band to keep their creative aspirations in perspective with the reality of the business – record sales.

On the other hand, when you're in the studio, you should give the band the benefit of the doubt whenever possible. Listen to their ideas and calculate how far off the deep end they are. In some cases you can let the record company play the bad guy.

If you feel so inclined, you can even act as a friendly consultant to the band, and get them to start thinking about long-term investments. Many young bucks can only see as far as the dollar signs in their eyes, and they are often grateful for such a long-term perspective. Of course, that is another book in itself.

As previously mentioned, the producer might also have to fight with the record company to preserve the artist's creative freedom. It is true that many executive producers and record company executives are completely out of touch with the aspects of a truly creative musical production. It is a good idea to search out any creative minds in the record company and ask them to help you with your creative campaign. Again, choose your battles wisely.

One way record companies intervene is by trying to clean up a project. A classic example was when they tried to get Aerosmith to stop drinking, doing drugs, and womanizing, which would've totally destroyed their creative process. Now they've at least given up the drinking and drugs, which were rampant in the rock-and-roll era. Elvis also was often told to clean things up. You can imagine what rappers have been going through since the beginning of hip-hop. Luke Skywalker, the rap artist, was involved in a huge battle over his lyrics to "Me So Horny." He was partially responsible for prompting the record industry to place parental advisory labels on albums.

On the other hand, through the years there have been record companies and executives who have taken big risks in the name of creative freedom, such as Berry Gordy, Jr. (Motown) and Herb Alpert (A&M). Clive Davis, who built CBS and Arista records, remains to this day a prime example of a top record company CEO who possesses the innate and majestic ability to recognize, cultivate, and market real talent. The vision of most record companies seems to have narrowed to merely a search for one-hit wonders, with little concern for the talent that creates a true lifetime superstar. When searching for a record company, you can still hope to find one run by an executive with the integrity of Clive Davis. Naturally, his goal is to make money; however, his reputation for looking after his artists is legendary, bar none!

Occasionally a band considered unacceptable by moneymaking standards will slip through the cracks and become a hit. But once the industry realizes it, they often market the hell out of the band, creating *serious* mass appeal. The trick is learning how to slip something past the record company's front line. Barring some type of fluke, the most common techniques are guerilla promotions and marketing. D'Angelo is one of the many artists who was successful with this approach. In addition, because of his tenacity he ultimately created a whole new style of music called neo-soul, which combines old-school R&B with hip-hop. If a band develops a large enough local following, with radio airplay and strong local record sales, record companies might very likely be enticed into signing, even if it doesn't fit what they think might be a normal hit.

Another technique is to simply figure out the limits of what the industry can handle and shoot for just beyond that edge. When you get in, keep pushing those limits bit by bit. Then, when you make it to the big time, do your part to help change the industry.

One school of thought that's still alive states that the best marketing you can do is to simply be yourself. Be the best you that you can be. Some people are simply touched by God. There are some quite famous recording engineers and producers who never listen to the radio. They have the creative genius within, and others follow them. However, they are extremely rare birds. Most producers need to be aware of market trends, especially because they *will* be dealing with the corporate structure of a record company.

Once again,

The beauty is that
pure connected creative genius
is
appealing to the masses.

Chapter 13: Moment-to-Moment Awareness

This chapter explains two techniques used to expand awareness while listening to, writing, or rearranging a song. Through the years, both of us have taught these techniques to our students. We have now realized that they share the same end result – moment-to-moment awareness.

This is what some might call an altered state of awareness, although for most producers it becomes as natural as can be. First, we'll explain what the state is like, and then we'll discuss each technique so you can experience it for yourself. Again, this level of awareness is critical for a professional producer, and it is a powerful technique for sharpening your writing and creative skills.

The State

When you are in this state of awareness, you are *in* the music. You have become the music. You are totally absorbed in all of the details, but you are also aware of the overall composition. It is similar to looking at a reflection of the sun on the water – you see hundreds of sparkles. You can look at any one sparkle, but occasionally an altered state kicks in, and all of a sudden you can see all of the sparkles at once, moment-to-moment. Then, when you really get into it, you start to see waves of patterns across the sparkles. Another example is watching the wind whip across a field of tall grass and noticing the overall patterns.

The analogy in music is that instead of focusing on one single detail of the arrangement or the mix, you *are* the mix and you not only hear every sound as it happens, but you also hear it in relation to all of the other sounds. Through the years when we have polled our students, we have found that more than 80 percent have had this type of experience, where they feel like they are *in* the music. They become the music. The trick is to be able to access this level of awareness at will – to be able to shift your awareness from focusing on specific details, to being in the music, then back to focusing on specific details.

When you are in the music, feeling every single note and its relationship to the whole, every little nuance is magnified. If something is awry, you know it – big time. Then, you shift your focus out of the overall awareness to the specific problem, and you fix it. Then, you dive back into the moment-to-moment awareness and see how it all feels again. Technically, when you are in this mode, your brain is in a state of Theta, which means your brain is mostly vibrating between 4 and 8 cycles per second. This Theta state is not a problem-solving mode; it is a feeling mode. Once you *feel* something is wrong, you then shift into problem-solving mode. After you fix the problem, you shift back into Theta mode.

In Theta, more than anything else, you are aware of timing from moment-to-moment because you hear every event as it happens. You also are unusually aware of frequency, or pitch. The frequency of every sound relative to every other sound is naturally evident as it occurs. You are also totally aware of the placement of each sound either between the speakers or around you – that is, the volume and panning relationships of each sound. But it is all focused in the presence of what's happening precisely in each moment.

It is like you are the center, the anchor. But the truth is, it is like you are invisible. In fact, we have noticed that when we are in this state of awareness, we are extremely still and quiet. You aren't scratching your leg or picking your nose, and you certainly aren't talking. You are totally in a state of awareness and stillness.

When you are listening to any production,
this type of awareness is absolutely critical.

David Gets into the Mix

I have visuals of mixes in real time. (You can see these on *The Art of Mixing* video series.) Each image of each sound flashes to the music based on the amplitude, the volume, and the dynamics of a musician's playing. I have shown these visuals to people all over the world in homes and at conferences and colleges. Even in high school, I remember being able to get *into* the mix. (This is probably why I became an engineer and a producer.) Once, while watching the visuals of the Pink Floyd song, "Time," I went into the altered state described earlier in this chapter. I found myself *inside* the mix – both aurally and visually. I realized that while I was in this state, that if there was anything in the mix that was wrong, I felt it throughout every cell of my body and psyche. I then realized that this was an incredibly powerful tool for a recording engineer. I later realized that all professional engineers and producers utilize this state of awareness, even though they might not be conscious of it.

I then asked myself, "How do you get in? What is the gateway to accessing this level of awareness?" I realized that if you focus on the low frequencies in the mix (normally the kick and the bass guitar) and figure out what they are doing, you can almost listen to them with your body. Then focus on the midrange sounds, including the left and right panning. Add these sounds to your perception of the bass sounds. Check out the highs (with panning) and add them in. You hear each sound as it happens, as though you're watching sparkles on the water. Before you know it, you're in!

Now that I have developed this ability to go in, I use it whenever I'm mixing or producing. After we get a basic mix developed, I go *in* and check it out. Then, I come back out and fix the volume, panning, EQ, or effects as necessary. Then, I go *back* in.

The key to this power is
knowing how to
control and focus your attention.

This is often referred to as developing a good ear. Most people think of a good ear as being able to distinguish between all of the sounds in the mix. This technique takes you a step further – to be able to hear all of the parts separately, or to be able to go into the mix and hear all of the parts in relation to each other at the same time.

Maestro's Composition of the Elements

During the late '70s and '80s I was vocal director and first piano chair for the university jazz orchestra, playing big band standards as well as my own compositions and arrangements. During this time, Frank Foster, who was running the Count Basie Orchestra, asked me to perform a piece I had written called "Bangle Bangle Snoot (The Belly Up Blues)," with the orchestra. When he asked me what the inspiration for the song was, it came to me that it was a "composition of the element." He validated this by saying, "That's heavy. You're going to be a world-famous composer one day." This really affected me, and I began to explore this inspiration. Soon after, I realized where the power in the composition of the elements lay.

I began to compose and write using the elements of nature as raw sounds – a car passing, a carpenter hammering, a bird chirping, or a conversation between two people, for example. Using a simple song format (ABAB form) as my platform, I put each of the sounds into a specific time signature (normally 4/4) and began visualizing a score emerging all around me *in real time*!

In this little experiment, every sound event was immediately visualized as a musical note with a particular tone, pitch, and rhythm. The mix of the sounds was based on how distant or near the sounds were.

Visual 101
Composition of the Elements in Nature

The key to the magic in this visualization exercise is to put each of the sounds into measures as they happen. The measures might be of equal length or they might be uneven, depending on whatever orchestration nature is giving you. I also realized that to get into the state of awareness in which the music was unfolding around me naturally, it was necessary for me to lie still quietly, with my eyes closed, and then focus on the sounds as they began to happen.

Sometimes the nature sounds would be transformed into common musical instruments. For example, the wind became the strings in the background. An automobile drawing near simulated the buildup of a cymbal rolling in the transition of a chord change in the song. Here is what it would look like in the studio in surround sound.

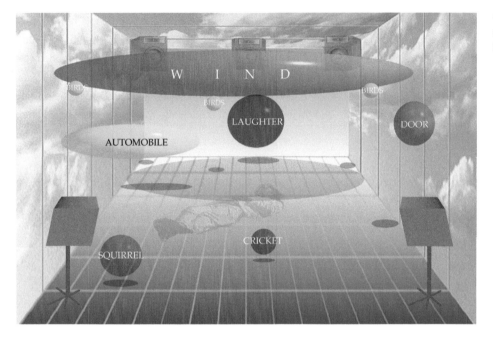

**Visual 102
Composition of the Elements
in Surround Sound**

Begin by deciding the number of measures that the song will have. I recommend starting with 24 to 32 bars. In one visualization, the song began with the chirping and singing of birds in the trees. This became the main melody of the song. I chose a 4/4 ballad for my time signature based on the initial rhythms of the nature sounds I heard.

The tempo was also determined by the birds (around 70 bpm). As you begin counting, everything in the world starts moving in time. Let everything around you fall into the measures effortlessly, just like a song. You may also use a metronome if one is available. As soon as you figure out the time signature, take a piece of paper and sketch out an empty 24-measure pattern completely separated with bar lines and a time signature, with multiple tracks, or *staves* (like a conductor's score).

You can use slashes, dots, or anything that you can draw quickly to represent the sound. Try to place them at the approximate pitches on the staff, and on their appropriate tracks. (You can hand-draw it.) (See Visual 103.)

**Visual 103
Composition of the Elements:
Conductor's Score**

In one sitting, the first eight measures were filled with the sounds of birds as the melody in the first two verses, which were four measures each. At the beginning of the second measure I could hear the faint sounds of laughter and voices from afar. I heard these sounds in the background of the mix.

Besides the birds and the distant laughter, the atmosphere was filled with many sounds. The wind subtlety came into the song like strings. The wind entered at the third measure. Meanwhile, the melody of the birds continued. The automobile cymbal sound started building at the beginning of the fourth measure. The more I focused, the keener my senses became, and I began to hear and feel many things at one time. For instance, by the time I got to the eighth measure I was tuned into my total environment – a grasshopper rattling leaves, a squirrel making a squeaking sound, a door in the distance opening and closing, the wind vibrating the leaves in the trees, the distant sound of voices talking, the birds chirping, a car that finally passed with its sound slowly fading into the background. As I listened, I continued counting and writing down each sound as it passed through each measure. At the ninth measure, the wind picked up and all of the sounds seemed to intensify. At this point I was simply *in* the mix of everything that was happening, and I noticed that most sounds seemed to get louder and climax together.

Whatever the case, it was a musical union of everything – the unity of every moment. The counting was like a mantra that connected me to everything around me.

I began to hear, feel, and sense all sounds in terms of musical measures. I began to hear the natural pitch of each sound, and how everything moves in rhythm and song mode – no different from music itself. In fact, there's absolutely no separation at all. It puts your spirit in tune with everything. What I discovered was that this process heightened my sense of feeling and hearing musical sounds to the point at which I could almost smell sounds. Composition of the elements exercises heighten one's ability to experience music and life from a truly organic perspective.

**The reward is
heightened sensory perception and
complete awareness of
all things around you.**

The key is to take this level of awareness into the production process. It is especially important when you are writing any musical parts for a production (or writing a whole song from scratch). It puts you into a multitrack awareness, aka "a multitrack mind." Amen.

**But most importantly,
it helps you to remember the music you come up with.**

Chapter 14: Songwriting

A songwriter's main objective is to touch the listening audience by painting as vivid a picture as possible with melody and words. It is like presenting a whole book in four minutes. Songs can take on a wide variety of forms. For example, you can use the hook in the intro to present the whole concept up front. Whether you start from a title or an idea, a song might tell a story just as if you were watching a movie or reading a book. On the other hand, some songs lead the listener through a maze, and the overall concept is not clear until the end. Other times, the song is a smattering of disconnected ideas with a theme sprinkled throughout that puts it all into perspective. Regardless of the structure, a song will always have a beginning, middle, and end. The songwriter often starts from an idea or title and builds the song to the very end.

Remember the saying, "Which came first, the chicken or the egg?" In songwriting there is a very similar question, "What comes first, the title, the lyrics, or the melody?" As previously covered, some songwriters start with a lyric, and then base the rest of the song on the inherent rhythm and musical content of the lyric. Others start with a melody. The melody, being the theme, will dictate the harmonies, rhythms, and even the mix. Some songwriters will actually start with a hook or a song title and develop the song around this central concept. The trick is to then go through each of the 11 aspects and incorporate them to fit the overall concept, melody, or lyrics. When all of these elements are fresh and combined with great harmonies, they make hit records.

Themes
Most professional songwriters write from the following themes:

- Love and fantasy
- Events and places
- Family and friends
- Realizations and self-reflections
- The latest dance or fad
- Heroes and legends
- Spirituality and religion
- Patriotism and politics
- The streets and survival
- Gimmicks
- Music itself.

Most songs focus on a main subject or theme. Some songs ask questions such as, "Is it the way you love me, baby?" or "When can I see you again?" These themes are easy to relate to for anyone who has dated someone whom they wanted to see again. Other songs simply make statements or declarations, such as, "Papa's got a brand new bag," or "I'm every woman." Other songs use titles of places, people, things, or businesses, such as "Route 66" and "MacArthur Park." Then there are songs that inspire you to act, including dance songs ("The Hustle" and "Shake, Rattle & Roll"), love songs ("Turn Off the Lights"), and inspirational songs ("Reach Out and Touch Somebody's Hand" and "We Are the World").

Songwriters also commonly use sub-themes as metaphors that add a deeper psychological effect. For instance, Aretha Franklin's "Dr. Feelgood" focuses on what a good lover her man is, and it is supported by several sub-themes – a jealous lover, being treated right, and being left alone by family and friends. You should try to be as creative as possible in your approach to using symbols and metaphors to tell your story.

Keep in mind there are times when less is best, particularly with dance songs, which have one main objective – to get you to let it all hang out. You shouldn't be bombarded with complex lyrics when you're trying to have fun. On the other hand, love songs can be as simple or as complex as you make them; however, you may want to keep in mind that 99.9 percent of hit love songs have universal (meaning anyone can relate) themes and are very simplistic in nature so listeners will have no problem feeling and understanding the song. A really good, solid song has a simple melody, a catchy hook, and a lyrical rhyme scheme that tells the story.

Some songs have structured rhyme schemes in which each word at the end of the phrase or sentence rhymes with the previous one. Sometimes rhymes happen in odd places, and other times there are no rhymes at all. Excessive rhyming can sometimes make a song sound corny. Of course, twisting the pronunciation of a word to make it rhyme can also sound silly. For example, Al Jolson had a song with the lyrics, "Mammy. This is your boy Sammy, from Alabammy. Please fix me some eggs and hammy."

ABAB Song Structures

Most popular music today is usually in ABAB (verse, chorus, verse, chorus) form and may include an intro and sometimes a bridge and lead break. There is also AABA (verse, verse, Bridge, verse) form. This form allows more time for a story to develop. Sly Stone cleverly used AABA in the song "Thank You" to tell a complete story before the chorus came.

If you have a song that has a strong enough hook (such as James Brown's rhythms, arrangement, and energy), the song structure and form could be negligible. Many hit songs have forms that technically are exceedingly simple, to the point of being boring. Sometimes it is the content that counts.

Using the Mix to Help the Song (The importance of Instrumentation)

Giving some forethought to the songwriting process, a good producer will help enhance the emotion in a song by highlighting certain instruments to create specific dynamics, thus supporting the songwriter's title, theme, and motive. Instruments can be highlighted by using many of the techniques covered in the 13 aspects or by bringing them out in the mix by turning up the volume slightly, adding short delays (called fattening), or brightening the EQ to make the sound more present.

Songwriting Help (Songwriting Organizations)

In most cities, songwriter associations such as NSAI are extremely helpful for helping you refine your songwriting craft. They can also be great places to make contacts to get your music placed. Once you've truly honed your songwriting craft, you can enter songs in songwriting workshops and shop them to publishing companies and record companies. You could even send your song directly to a major producer or artist.

Most songwriter associations, publishers, and record companies have websites that you can check out for more specific information. Songwriter.com is a website that is extremely useful and highly recommended by some of today's producers.

PART 7

Selecting or Becoming a Producer

Chapter 15: Choosing a Producer

This chapter is designed for someone who wants to hire a producer, but it will also help you, as a producer, to see what people are looking for (or should be looking for) so you can begin molding your skill sets to fit the market.

What to Look for in a Producer

When you are choosing a producer, look for two things – someone who has the experience and skills to get the job done and someone whose values are similar to yours. In fact, depending on your skill level, you are looking for a producer who can teach you what is cool, even if it's by osmosis.

The problem is that there are a huge number of producers who are flying by the seat of some very thin pants. After reading this book, if you come across a producer who barely knows what he or she is doing, you will know. Be careful, though. The precise techniques for producing have never been established (until this book), and there might be many great producers out there who are lacking some important skills explained herein.

Most people think that a great producer is someone who has a lot of hits to his or her credit. In one respect, it is hard to argue with a hit. However, if you can't get along with a particular producer, then he or she is not so great for you. You need a producer who will relate to you at the level you are at, and who will give you the respect you deserve.

As previously mentioned, a great producer has the following skills:

* Good listening skills
* The ability to organize a large amount of diverse information
* The ability to facilitate and work with people
* An understanding of a producer's job and the producing process.

It is also helpful (but not absolutely necessary) for a great producer to have:

* A thorough background as a musician
* A strong background in music theory
* Knowledge of the techniques of other producers.

Finding a Producer

Before you start looking for a producer, make a checklist of the traits that are the most important to you. You also need to figure out how much musical help you might need. Do you need a music producer, or would an engineering producer be sufficient? If you are a good enough musician, you very well might not need a music producer. In fact, one could conceivably slow you down.

Major record companies commonly will choose a producer for you, unless, when you negotiate the record deal, they agree to mutually select the producer. If you have already been working with a producer who you really like, you can sometimes talk the record company into using that person.

If you have already been checking out producers, you might have a list of a few you would like to use. If not, talk to other bands in your area for a referral. Sometimes you can even call local studios to get names and numbers. If you have a larger budget or are working with a record company, look at who produced some of your favorite albums.

You can also find lists of music producers on the Web and in the *Recording Industry Sourcebook* (ArtistPro, 2004). Management companies can also help. For example, the Terry Lippman Company in West Hollywood, California, keeps its own list of producers.

How to Interview a Producer

You can do an initial phone interview with your potential producer and ask him the following questions before he has heard your music:

- "Who have you produced? Can you send us copies of your best work?"
- "What makes you a great producer?" This one always gets them going. It is an especially interesting question because you might get a defensive answer from those who are insecure or have a big ego. Listen closely to whether the producer gets defensive because it can be a good clue as to how he or she might react to your suggestions or ideas later on in the project.
- "What are your weaknesses?" This one also has the potential to reveal some deep-seated issues. Again, you can often learn more from the producer's attitude when he or she responds than from what is actually said.
- "What is your favorite music?"
- "What makes a great production?"
- "Do you follow a typical production process?"
- "What makes a good mix?"

Because these questions are so vague, they could easily upset or frustrate an inexperienced producer. The trick is to ask very difficult questions to see how the producer reacts when stressed.

If the phone interview goes well, send him or her a CD of your material, and then meet for a more in-depth interview. Here is a list of questions to ask a potential producer after he or she has heard your music:

- "What do you think?" It's always interesting to see the context in which they put their answers. Some might say, "Cool," and some might say, "Good here, needs work there." Listen to how the producer approaches it and how detailed he or she gets.

- "How would you characterize our music?" The way the producer answers can be very telling. You should be quite forgiving if the producer has a different idea at first, before you have explained your vision for the album.
- "Do you think it should be recorded live, with minimal overdubs, or with lots of layering?"
- "What are some examples of some techniques you might employ to make it a better project?"

How to Present Yourself to a Producer

If a producer doesn't perceive that you have what it takes to make a hit (or that you have a huge amount of money to throw down), he or she very likely will not take on your project. Producers are looking for a project that has excitement, enthusiasm, and energy. But most important, they are looking for great songs, great talent, and mature individuals who are capable of becoming major artists. Of course, they would also love to work with a band that already has a record deal. The goal is to get the best producer possible, based on your level of expertise. You must demonstrate a certain level of credibility, savvy, and an awareness of the industry. Do your homework. It is important for you to be able to converse about your project, and to have a perspective on what you are trying to achieve (other than just a hit).

After reading this book, you will know how to critique your own project and get it to the next level. You also will know what it takes to take your project all the way. What the hell – produce it yourself. Having read this book, you know precisely what the job entails.

Chapter 16: Selling Yourself as a Producer

Once you've finished producing your first band, they will know big time that you have made a humongous difference in the production and recording (especially if they have recorded in the studio before). You made sure the concept was cohesive and got everyone to come up with ideas that made it stronger. You helped examine all of the musical parts to make sure they were the best they could be. The melody is appropriate, the rhythm fits perfectly, and the harmonies are arranged correctly. You've got lyrics that have been examined and refined, and are now perfect. You've made sure the song structure is the best it can be. You've got an arrangement with layers that build up and break down, creating some intense dynamics. You've got instruments and instrument sounds that are really cool, and there are no problems with any of the instruments. You've got band members who (hopefully) are taking lessons. You had the mix planned out and preconceived. And, you chose the appropriate studio to record in, and made sure that you had all the right equipment for the recording session.

Most important,
you found the hooks and
brought out the essence of the song.

It will be obvious to anyone that you have made a huge difference. That's when your band tells other bands, "You're going into the studio to record . . . Hey man, check out this guy. He made a huge difference in the recording, and it was really cool." This is the number one way to get hired as a producer.

One of our students produced a band for free for the first time. After they started getting airplay on the radio, other bands started asking them who produced their project. We started getting calls from bands looking for this student to produce them. Referrals are your number one way to get hired as a producer. Therefore, start producing as many bands as you can, regardless of whether you get paid! Then you are on your way to establishing a reputation and getting more jobs as a result of word of mouth.

Getting the Job: Finding Bands

If you just go to a club, tell bands that you are a producer, and ask them whether they want you to produce them, you're likely to get a response like, "Who the hell are *you*?" As previously discussed, there is too much fear of producer running rampant through our society. And no one is going to hire you unless you can demonstrate some credibility.

One technique is to go to a band that you like and say, "I *really* like your music and think you've got a lot of potential. I've produced a lot of bands and I've got this technique that I take people through to produce a band. What I would like to do is come to your rehearsal and show you this technique and see what you guys think. What do you think?" Because you are not asking them to commit in advance, most bands are open to the idea. Then, when you get into the rehearsal, tell them about your procedures for critiquing each of the 13 aspects, but emphasize that you are very careful to help them bring out their own ideas and create their own sound. You are not there to tell them what to do. This is one proven technique for introducing yourself to bands.

If you are a recording engineer, you may try to negotiate points (record sales percentages) on a project as a producer in lieu of payment or partial payment. Many times, this is how an engineer gets his or her foot in the door as a producer. If you can afford the time, this can be a very smart way to develop a reputation and a following. If the album then becomes a hit, you've got it made.

If you own a recording studio, you can use studio time as a way to bring in a project. Obviously, if you offer free studio time (or even just a discount), you look much more attractive as a producer. But don't sell yourself short either. If you are trying to sell yourself as a producer and you don't want to give the band free studio time, you must let them know as early as possible that you are not offering free studio time in the deal. Many bands tend to initially assume (or hope intensely) that you will be giving them the studio time.

Poop Sheets

You need to put together a poop sheet (yes, that's a technical term) – a little marketing tool – on who you are and how you can help people as a producer. Explain in brief the process through which you will take the band. Do it in color. Then you are a real producer.

Rate Sheets

Once you have your poop sheet, decide what you are going to charge. Gauge it based on everyone's perception of your expertise (including your own) and your perception of how much the band can afford. A small-time producer might charge $2,000 up front, and then $2,000 upon completion of the album. You might also charge a small hourly fee in case the project goes on forever. We know one producer who used to charge an additional $10 per hour while in the studio. If you are a recording engineer, you might add another $15 to $30 per hour to your engineering fees. In exchange for cash, you may negotiate royalties on the project. The percentage of royalties is called *points*. The normal rate might be four points or 4 percent (which is 4 percent *after* the record company takes its approximate 85 percent). At this point it is important to talk to an attorney and have him or her draw up a contract.

One way to proceed on your second project is tell the band that you aren't going to charge them anything for the project. But, if they feel at the end of the project that you've made a *huge* difference, then ask for, say, $500 to $2,000. Even if they end up not paying you at all, you have now produced *two* projects. You now have two bands that will very likely give referrals, which can lead to actual paying gigs. Even if you end up doing five projects for free, you've become much more experienced at your craft. And when you approach a band and say, "I really know what I'm doing. I've produced a lot of bands, and I can *really* help you," that air of confidence will comes across and they will say, "Okay." Then you are on your way.

Unsigned artists don't necessarily have to give royalties to a producer. Record companies typically pay a mid-level producer $50,000 per album ($8,000 advance), or $8,000 for a single. Major producers obviously get more. If Bob Ezrin were to produce the next Pink Floyd album, the record label would give him a sizeable advance before he even went into the studio. On the last Pink Floyd album he produced, it is likely that he received at least seven figures! And you know Quincy Jones is set. He has probably made more than $10,000,000 just from the *Thriller* album alone, and the cheese just keeps coming. And I don't mean Limburger!

PART 8

Protecting Your Music

Additional contributions by Jon L. Duman, Esq.

Chapter 17: Protection

Having a complete understanding of the music business includes the legal and protection end as well as the production and performing aspects. This means if you are the writer, author, or composer of a song, there are specific procedures you must adhere to in order to protect yourself from wrongdoing by others against you and your music. Theft in this business is as common as fraud and larceny anywhere else in the world. Protecting your songs is extremely important and is not to be taken lightly. People can capitalize and make bazillions off your creative work. Don't be a victim! We've seen this happen to friends, and it's not a pretty sight!

Copyrights

Under U.S. Copyright law, your ownership of a song (the "copyright") begins as soon as you finish writing the song. As long as he or she has not entered into a written agreement which gives the copyright to someone else, the creator of the song is presumed to be the owner of the copyright. It may become necessary, however, for the owner of a song to show proof of ownership, including when ownership began, in the event that someone challenges or illegally infringes upon, the owner's rights in the song. One method to establish proof of ownership is the so-called "poor man's copyright," which consists of sending a copy of the song to yourself in the mail. The postmarked date on the mailed package can thereafter be used as evidence of when your ownership rights began. In order for this method to work, you cannot open the package or break the seal. The package should thereafter be kept it in a safe place for further verification should you need it in the future. It is important to know that the "poor man's copyright" may not be effective to prove ownership in all cases. There have been court cases where it has not held up in court. However, it is, in fact, "better than nothing."

The more preferable option is to formally register the song with the Library of Congress in Washington, D.C. Among other benefits, formally registering your music ensures your right to receive statutory damages of up to $150,000 per infringement, should someone steal or plagiarize your music. There are two types of forms used to register musical works: Form PA (Performing Arts) and Form SR (Sound Recording). You use the PA form if you only want or need to copyright the actual composition and not the recording. The SR form copyrights the actual recording, including the performance and the production elements of the sound recording. You can use the SR form to protect both your composition and the recording, provided that you are the owner, in equal percentages, of both. If you record someone else's song, you can actually still copyright your own recording of it (although you still need to get the other owner's permission to commercially exploit anything containing that owner's material).

We have included references to the copyright forms in Appendix D, "Forms"; however, you can easily download them from the Web at www.loc.gov/copyright. You can also request forms by writing to the Library of Congress, Copyright Office, 101 Independence Ave. S.E., Washington, D.C., 20559–6000. It doesn't take a rocket scientist to copyright your work. The forms are very simple to fill out. Enclose a tape, CD, *or* sheet music. The number of recordings you submit is based on the following criteria:

- If the work is unpublished, send one tape or disc consisting of a clear recording of the song.
- If it was first published in the United States on or after January 1, 1978, send two complete tapes or discs of the best edition of the work.
- For work first published outside the United States after March 1, 1989, send one complete tape or disc of either the first published edition or the best edition of the work.

Note that your recordings will not be returned. Send the required $30 fee for registration payable to Register of Copyrights. You don't need to include the sheet music or even the lyrics – just the tape or CD. You can copyright as many songs at once as you like, as a collection of work, as long as the ownership of all of the songs is the same. There are, however, practical advantages of registering songs individually, as opposed to as a collection, because among other things, songs that are grouped and registered together are not necessarily listed individually in the official record.

Although you can also register your songs with agencies such as ASCAP, BMI, and SESAC, registration with these companies is not a substitute for formally registering the copyright with the Library of Congress.

Copyright Income

There are four main sources of potential payment available to you from the exploitation of your music: (a) record sales and licensing income; (b) mechanical royalties; (c) performance royalties; and (d) synchronization income (see below). Record sales and licensing income is paid to the owner of the copyrights. This may be the record company, artist, or the band. Of course, if you have a record deal then the record company pays you (after they have taken out payments for certain overheads listed below). The rest of royalties: mechanical, performance, and synchronization are paid to the songwriter, who owns the copyright for the songs (not the recording). Sometimes the songwriter will assign part of their royalties to a publisher to aid in placing their songs with other artists or in films, for example. These publishing rights are commonly 50 percent of the songwriter's royalties, but may be negotiated at any percentage desired. The songwriter owns all publishing rights from the moment they write the song.

Record Sales and Licensing Income

This is the money that is paid to the owners of the copyright to a master recording. This category of income will only apply to you if you are either the owner of, or a featured performer in, master recordings. In situations where a songwriter/recording artist retains ownership of his or her own recordings, income from the sales or licensing of the recordings are paid directly to the artist by distributors, wholesalers, and retail outlets. More commonly, however, a record label may own the artist's recordings, in which case all income from the sale or license of the recordings is paid to the label, who will make various "standard" industry adjustments (including deductions for recording and promotional costs, third party mechanical royalty payment, tour support, "free" goods, and packaging, among others), before paying the featured artist or band a small percentage of the remainder (referred to as the "artist's royalty"). Depending upon your bargaining position, the standard artist's royalty rate ranges from between 10 to 20 percent (or "points") of the adjusted profits from the sale of records, which, after payment of the producer's share, can yield up to between $1 to $2 per album for the artist or band. This may not seem like a lot, but multiply that by a million for a platinum (million seller) album, and you or your band would get $1 to $2 million dollars! Then, once again, it doesn't seem like a lot when you realize the record company is getting the other $8 million or so.

Mechanical License Royalties

These royalties are paid to the songwriter or owner of the copyright to a song (or his or her publisher), in exchange for the licensed permission to make and sell sound (mechanical) recordings featuring the song. The money paid to the songwriter comes from actual record sales. Unless the songwriter or publisher agrees to a different arrangement, the party seeking to record the song (e.g. the record label) must pay the owner a mechanical royalty which is set by statute (currently 8.5 cents), for each copy of the recorded song which is distributed for sale. Where an entire album is released, the 8.5 cent rate is multiplied by the number of recorded songs on the album (e.g. 10 songs would require a total mechanical royalty payment of 85 cents per album). Where a single recorded song has more than one owner, they will split the 8.5 cent mechanical royalty between them. Over 80 percent of mechanical royalties paid in the United States are collected by the Harry Fox Agency, a company that collects these royalties from the record labels and other companies who owe them, and, after deducting a percentage commission off the top as its fee (around 4 percent), remits the difference to the owner or publisher of the song. Note that it is 4 percent of the 8.5 cents per song that Harry Fox keeps (not 4 percent of the record sales).

Public Performance Royalties

Performance royalties are also paid to the songwriter and/or the publisher of a song, as a result of the public performance or public broadcast of performances of the song. Most public performance royalties come from the use of songs as part of television and radio broadcasts. However, you are entitled to receive a public performance royalty whenever your music gets performed anywhere for financial gain – including when it is played in a club or a clothing store. ASCAP, BMI, and SESAC are the three main agencies within the U.S. which administer public performance rights, including the collection and distribution of performance royalties due to the song owners. These companies regularly monitor broadcast logs submitted by radio and TV stations, and apply a statistical formula in order to determine the amount of performance royalties due from each user with regard to a particular song.

Here is a list of users who regularly pay public performance royalties to ASCAP, BMI, and SESAC:

- Major television networks: ABC, CBS, and NBC
- The Public Broadcasting System (PBS) and its affiliated stations
- The majority of the 11,000 cable systems and virtually all of the cable program services
- More than 1,000 local commercial television stations, including affiliates of Fox, Paramount (UPN), Warner Bros. (WB), and PAX networks
- The Univision Television Network and its stations
- Approximately 11,500 local commercial radio stations
- Approximately 2,000 non-commercial radio broadcasters, including college radio stations and National Public Radio (NPR) stations
- Hundreds of background music services (such as Muzak and airlines)
  - Approximately 2,300 colleges and universities
  - Approximately 5,700 concert presenters
  - More than 1,000 symphony orchestras
  - More than 20,000 websites
- Tens of thousands of "general" licensees – bars, restaurants, hotels, ice- and roller-skating rinks, circuses, theme parks, veterans' and fraternal organizations, and more.

As is the case with the Harry Fox Agency, ASCAP, BMI, and SESAC will deduct a percentage-based fee from collected royalties, before distributing the remainder to the songwriter and publisher of the song. Payments of public performance royalties from these agencies are split between the publisher (who receives the "publisher's share"), and the songwriter (who receives the "writer's share"). A songwriter who acts as his or her own publisher will receive both shares. Public performance royalties can potentially result in a substantial amount of money, totaling several hundred thousand dollars a year for a single "hit" song. An artist who has multiple hits over many years can do quite well for the rest of their life, solely on public performance royalties.

Synchronization Income . .

Synchronization Income is paid to the songwriter or owner of a song (or his or her publisher), in exchange for granting permission to use the song in combination with some sort of visual image. "Synch Licenses" are commonly negotiated for the use of music in televised commercials, film, videos, television broadcasts, video web-casts, or other audiovisual productions. Synchronization fees are negotiated by contract and vary according to the popularity of the song, as well as the prominence of the song's use within the audiovisual work. For example, the synchronization fee to use an entire song in a major motion picture might start in the $50,000 to $75,000 range (upwards to several hundred thousand dollars for certain "hit" songs), whereas use of only a portion of a relatively unknown song as background music in a daytime television show might range from $1,000 to $2,000. Synchronization income is paid directly by the licensing party to the owner or publisher of the song.

The following chart shows the typical split of monies between the owner of the band versus the songwriter/publisher. This example is when a record company is involved. Of course, if you are contracting your own distribution then there are no deductions made by a record company.

Money Paid to the Artist or Owners of the Band	Money Paid to the Songwriter / Publisher
1. Record Sales and Licensing Income **2. Mechanical Royalties** 8.5¢ per song (85¢ per CD with 10 songs) goes to the Songwriter/Publisher (Collected by Harry Fox Agency) Approximately 4% retained by the Fox Agency for collecting the money *Deductions made by the Record Company:* Standard Free Goods – Promos, review & distributor copies Special Free Goods – Discount Programs Packaging Deductions **Reserves** **Artwork** The Artist/Band gets 10–20% of what's left. The Record Company gets the rest.	Typically the split is as follows: – Songwriter gets 50% – Publisher gets 50% **2. Mechanical Royalties** (8.8¢ per song) **3. Performance Royalties** (Collected by ASCAP, BMI, SESAC) **Soundtracks.** **Small Performing Rights** – Radio, TV, Juke Boxes. **Grand Performing Rights** – Musical theatre pieces. **Print rights** – Lyrics **4. Synchronization Income** Paid by anyone who uses the song as part of film, TV, video, etc.

Chart 11 Royalties Chart

Publishers

When you write a song, you are initially both the songwriter and the publisher, meaning that you normally retain all of the publishing rights in your song, unless you elect to transfer ownership to some or all of them. For example, you can elect to give up your publishing rights to a publishing company, in which case your publisher will take care of all of the paperwork involving your music publishing, and will use its network of connections to get your song placed on other people's albums and in commercials, videos, or movies. Although it is possible to administer your own publishing, only major songwriters commonly have enough resources to do so effectively. In the event that you decide to transfer your publishing rights to a publisher or other company, U.S. copyright law now provides that your publishing rights may revert to you after 35 years. It's also possible to negotiate a reversion clause in your publishing record deal which provides for a reversion of rights in the event that the publisher fails to generate a minimum level of income or activity for your song(s).

Established songwriters can also sometimes negotiate a substantial cash advance from his or her publishing company, which is recoupable against future publishing royalties.

Standard publishing agreements provide that the publisher collects all mechanical royalties, synchronization income, and any other publishing income (except for the "writer's share" of public performance royalties, which ASCAP, BMI, and/or SESAC will send directly to the songwriter). The publisher will then deduct its expenses, as well as its agreed-upon share (typically 50 percent), from the publishing income it collects, before sending the remainder to the songwriter.

It is only recommended to allow a larger publishing company to handle your publishing, such as MCA Music, Warner/Chappell, BMG Music, and Sony Music (all of which are divisions of the industry's largest record labels). Although many smaller record companies also have their own publishing divisions, you shouldn't agree to let your record label also act as your music publisher, unless the label can demonstrate that it has the resources and experience necessary to effectively administer your publishing rights.

It should be noted that many songwriters may not necessarily be interested in recording their own material. They can make a good living off of mechanicals and performance royalties and synch income, simply by letting a publishing company place their music for use by other artists or companies.

Personal (Artist) Managers

A manager is almost considered to be a part of the band when it comes to a record deal. Occasionally, he or she will also get points in a record deal. Personal Management agreements usually require the artist to work only with a particular manager for a period of 3–5 years. However, artists should always retain the final say regarding the approval of contracts and should feel free to always question any decisions that a manager makes without them.

A manager's compensation is typically based on a percentage (ranging between 15–25 percent) of the artist's overall income from the entertainment industries, which can include publishing income received from the artist's activities as a songwriter. However, depending upon your bargaining strength, you may be able to negotiate the exclusion of music publishing income, or certain other types of income, from calculation of the manager's commission. An established artist can also negotiate conditions under which he or she may terminate the management contract prematurely (such as where the manager fails to secure a record deal within a designated period of time).

Negotiating Record Deals

This section provides a list of items that are negotiated in a normal record deal. However, it is highly recommended that you work with an attorney or an experienced negotiator on a record deal. You can often find an attorney from your local Bar association. Some states have Lawyer for the Arts groups that can refer you to an entertainment attorney (for example, the California Lawyers for the Arts maintains offices in San Francisco, Oakland, Santa Monica, and Sacramento). You also can find attorneys at the Music Business Attorney Legal and Business Affairs Registry (www.musicregistry.com), or at NOLO Law for All (www.nolopress.com – search for music law).

All of the following items are on the table when you are negotiating record deals. It is a good idea to prioritize them with your lawyer before negotiating with the label.

- Number of albums. A "new" artist or band will generally be obligated to complete a larger number of albums ("the delivery commitment") than an "established" artist.
- Recording budget and advance. Typical amounts can range from $50,000 to 125,000 for the first album (which will be expected to cover your recording budget, advance, manager's commission, and producer fees) with appropriate increases of 25 percent or more for subsequent albums.
- Royalties. Twelve points is good; 15 points is very good; 20 points is rare. The artist or band must also negotiate how many points they are willing to give the producer. Three points is normal, four to five points if they are "big time." The artist or band is responsible to pay the producer out their own royalties. Record companies rarely use in-house producers these days.
- Amount of escalating royalties. If sales are big, up to 20 points (20 percent).
- Standard free goods. This is an example of the types of deductions a record label makes from records sales, before calculating the artist's royalty. Other examples include packaging, breakage, and "new technology" deductions. Examples of "standard free goods" include promo copies, review copies, and distributor copies. Although this is commonly non-negotiable for new artists and groups, you should try to negotiate as small a number as possible (since these are records that you are not paid a royalty for).
- Special free goods. These include discount programs for retailers who stock your album (see comments above).
- Royalties in other countries. These are typically based on a lesser percentage that the royalty rate paid for domestic sales. The following suggestions will give you a guide:
  - Canada: 85 Percent
  - Key Territories (Uk, Germany, Switzerland, Scandinavia, Italy, Australia): 75 Percent
  - France, Japan, And Other Countries: 60 Percent
- Escalations In Key Territories. This Is Only Important If The Band Has Had Previous European Album Releases.
- Packaging Deductions For Cds. This Is Another Example Of The Types Of Deductions A Record Label Makes From Records Sales, Before Calculating The Artist's Royalty. The Following Numbers Should Give You A Guide:
  - Vinyl/Sleeves: No More Than 10 Percent
  - Double Albums: No More Than 15 Percent
  - Cds/Digipaks: no more than 25 percent
- Recoupment of mechanicals. Ordinarily, the label is not supposed to deduct (recoup) from the artist's royalties those mechanical license fees which the label pays to the artist for use of the artist's songs on the artist's own recordings. However, mechanical fees paid to songwriters other than the artist may be deducted (recouped) from the artist's royalty.
- Mechanical rates. Most recording contracts provide that the label only has to pay the artist a mechanical license fee of only 75 percent of the standard 8.5 cent mechanical rate for use of the artist's songs on the artist's own recordings. Similarly, the label will usually pay for maximum of 10 songs (even if the artist contributes 20 songs!).
- Reserves. This refers to the practice of withholding a portion of the Artist's royalty as insurance against returned CDs. How quickly the reserves are liquidated if the CD becomes a hit is negotiable (between 2–4 accounting periods is standard).

- Video production costs. Record companies will typically advance payment for music video production costs. Normally 50 percent of the production cost is recoupable. Two guaranteed videos is rare. The video production budget may be determined after the recording contract is signed.

- Independent promotion. Record companies will also typically advance payment for independent promotion services. Normally 50 percent of the cost of independent promotion is recoupable.

- Tour support. Where provided by the label, this is almost always totally recoupable (including from other sources of income due to the artist – a practice known as "cross-collateralization"). For a four-man, eight-week tour at $25,000–30,000 per week, an artist or group normally might make $5,000 per week; therefore, the cost at $25,000–30,000 per week for eight weeks would be $160,000–200,000.

- Release commitment. Normally, the label is supposed to release an album within 90 to 180 days after delivery of the completed album. You should be able to negotiate for the right to terminate the agreement if this doesn't happen. You may also be able to negotiate the right to purchase the unreleased recordings and release them yourself, by paying the label for the cost of recording the album (otherwise, record companies are not required to give the rights to the recordings back to the artist).

- Artwork. For album artwork contributed by a well-known (in demand) graphic artist, the creator will typically retain ownership of the copyright in the artwork and a royalty may be payable to the creator for use of the artwork. Amounts paid by the label for this purpose is generally recoupable.

- Group breakup clause (also known as a "Leaving member clause"). If a band member quits, or if the group completely disbands, the individual members will usually be contractually obligated, upon demand, to sign separate recording contracts with the label. Where possible, artists may want to negotiate a minimum recording budget for the leaving members' subsequent projects with the label. In addition, procedures for recording a demo for solo departing members may also be considered.

Appendix A: Checklist of All Homework to Be Done before the Pre-Production Meeting

1 Keep a notebook for each project.
2 Get the songs on tape.
3 Get the lyrics on paper. Make copies of original lyric sheet and one with all your notes on it for everyone.
4 Make a song map. Make copies of the original song map and one with all your notes on it for everyone.
5 Schedule the order of the recording of the songs and instruments.

Aspect 1: Concept

1 Figure out what the concept is and write it down.
2 Look to see if any of the 11 aspects support the concept. How strong is the component? Write it down.
3 Think of things that add to the concept. Can a supporting aspect be made stronger?
4 Critique the name of the song. See whether you like it and it fits. If not, try to come up with some ideas for a different name.

Aspect 2: Intention

1 See if you can come up with a short intention based on the concept.
2 Practice meditating on this intention yourself.

Aspect 3: Melody

1 Note where the key melodies happen in the song and make a note in your song map. Note what instruments are playing the melodies.
2 Write down your first impressions of the melody. How does it strike you? How does it feel? What might you compare it to?
3 How central is each melody to the song? What is goal of the melody?
4 Does the melody fit the song? Is it appropriate? Does it support the concept of the song?
5 Is the melody too simple or complex (busy)?
6 If you have a melody line that is a major hook in the song, look for a place where it might be repeated. Or, think about placing it low in the mix in another section of the song. Or, you might use another instrument or part to hint at it.

7   Note if there are any sub-melodies or counter-melodies. Could the melody be reinforced with sub-melodies and counter-melodies? Are there any places where you might add a lead-in?

8   How predicable or unpredictable is the melody? Do you like it the way it is and does it fit the song? Might it be appropriate to consider having the melody change from being unpredictable to predictable (or vice versa)?

9   Is the melody being played or sung "straight" the first time it is introduced in the chorus?

10   Does it remind you of another song? Is it already used precisely in another song? If so, you need to bring up the possibility of copyright infringement with the band.

Aspect 4: Rhythm

1   Check to see whether the type of rhythm fits the style of music. Does it feel right?

2   Listen to see whether the instrument selection is right for the rhythm. Think about whether you might choose other instruments.

3   Figure out the exact tempo in beats per minute (bpm).

4   Critique the tempo.

5   Critique whether the rhythm is too busy or too simple. Look to see whether the rhythm should get more complex or be more simple in any sections of the song.

6   See whether you would like to add any polyrhythms to spice it up. Note the accents and see whether it might be appropriate to change them. Consider adding pickup notes and grace notes.

7   Make sure no sound is walking on any other sound rhythmically. Look for parts that are too busy.

8   Listen for any places where you might divvy up the rhythm among instruments. If they exist, think about how they might be panned or effected differently.

9   Explore using different sounds for traditional parts, such as kick, snare, and hi-hat.

Aspect 5: Harmony

1   On your lyric sheet, mark where each harmony part will be. Do they add to the song? Where else might it be appropriate to add harmonies?

2   Decide on the chord structure and voicing of the harmonies.

3   Decide how many parts the harmonies will consist of – unison, one part, two parts, three parts, and so on.

4   Decide how many layers and tracks the harmonies will be recorded on.

5   Decide how the harmonies will be mixed as far as volume, panning, EQ, and effects.

6   Consider creating different textures by treating harmonies in different sections of the song differently.

Aspect 6: Lyrics

Write down your comments or issues on the lyrics sheets.

1   See whether each line supports the concept or diverges too much.

2   Listen for the level of magic (or lack thereof) in the way the lyrics synchronize with the music.

3   Are the lyrics rhythmically correct? If not, how bad are they?

4   Critique the lyrical phrasing on every line.

5   Look for any enunciation problems that really bother you.

6   Ask yourself whether any word or line in the lyrics could be taken out without affecting the meaning. If so, see whether it is possible musically to take out the words.

7   Listen for any words that are clichéd, overtly offensive, or outright evil. Make your judgment about whether you want to fight that battle.

Aspect 7: Density of Arrangement

1   Ask yourself whether there should be more or fewer sounds at any moment. Consider double- or triple-tracking or adding parts.

2   Should you have more or fewer notes, or more sustaining, legato sounds?

3   Should you have more effects filling in the mix?

4   Figure out an overall progression from simple to complex (and vice versa) that fits the song.

5   Take a look at the density within each frequency range. Is it cluttered in any range? Is it spread evenly? Is it the way you would like it?

6   Figure out how many tracks you will need. Budget your tracks, if necessary.

7   Make copies of the original arrangement the band started with, but also make copies of the version with all of your ideas added.

Aspect 8: Instrumentation

1   Decide whether you would like to use common instrumentations for the style of music with which you are working. If so, think about where you might replace sounds and with what instruments.

2   Do the sounds fit the overall song and its concept?

3   Do the sounds work well together? Conceptualize your overall blend or separateness of sounds throughout the song.

4   Are the sounds of the highest quality with no problems?

5   Try to come up with a signature sound? Also, come up with unique sounds that you could place low in the mix.

6   Have the band create their own sounds with the synthesizer and/or find their own samples out in the world.

Aspect 9: Song Structure

1   Listen to see whether you might add, take out, or shorten a section.

2   Ask yourself whether any section doesn't feel right. If it doesn't, figure out why and see what you can do to fix it.

3   Does every section of the song flow smoothly into the next section? Think of different ways to smooth over the transitions.

Aspect 10: Performance

Do this homework at a rehearsal and again while listening to the demo tape.

1   Note the level of expertise for each player. Look for any major weaknesses.

2   Make sure the band is well rehearsed, all MIDI parts are sequenced, and all parts are worked out.

3   Listen for any difficult sections in which it is obvious that a musician is barely keeping up. Note these spots on your song map.

Aspect 11: Mix

1   Design the overall mix that is appropriate for each song.

2   On your lyric sheet, map out all the places where you might want to have a delay on the end of a line.

Aspect 12: Quality of the Equipment and the Recording

1   Make a list of all the high-end equipment you will need for the session.

2   Go to the studio where you will be recording and see whether they have all the equipment you need.

3   Check out the size of the rooms and the number of isolation booths.

4   Check out the studio comforts, such as TV, videos, and games. Often it nice to have these perks to keep musicians from being too bored while they're waiting to play. Of course, these are not nearly as important as sound, price, and equipment.

5   Make sure that all of the equipment at the studio is in good working order.

6   Listen to how the studio responds to your requests and concerns.

7   Before booking the studio time, collect the schedules and availability of each band member. Call the studio and schedule each of the sessions. Confirm the schedules with each musician.

Aspect 13: Hooks

1   Define all of the hooks in the song. Determine which are the strongest and most important for the overall song.

2   Determine which one of the 11 aspects contains each of the hooks.

3   Is the hook at the beginning, and do you want it to be? You might start with the hook on one song to keep the album flow interesting.

4   Think about how you might repeat or use variations of the hooks in the song.

Creating a Style of Music

1   Determine what the primary style of music is. List any other influences. Think about the overall project and how it might best be marketed.

2   Go through each of the 11 aspects and write down what it is in each aspect that determines the style of music. Think about whether you would like to adjust that component to play down or highlight the style that is coming through the particular aspect.

Order of the Songs

1   Figure out the order of the songs. Write down the reasons for your decisions. Make CDs with various orders.

Appendix B: Nomenclature and Monosyllables for Describing Musical and Rhythmical Patterns

These terms can be especially helpful when you are trying to sing to musicians how a part should go.

General Instrument Syallables

- **Kick.** Boom, Koonk, Hoomph.
- **Snare.** Che, Te, Tak, To. A kick snare part would be Koonk, Tak, Koonk, Koonk, Tak.
- **Congas and toms.** Duum. Common lick might be Pa-Dooka, Dooka, Dak.
- **Hi-hat.** Open and closed pattern might sound like Sche, Tse, Tse, Sche.
- **Bass.** B, B, Bumb, B, B, Bumb.
- **Horns.** Waddup (Wah, Da, Du, Dah, Da, Du, Dah). Dwee, Doodle, Dattle, Do, Dwee.
- **Guitar.** Kechink or Kechank. For example, Kechinka, Chank, Kechank.
- Distorted Guitar. Weh-Ou.

Tabla Sounds (Bol)

• Dha (as in "Ad-hoc")	• Na (as in "Nut")
• Dhin (as in "Hinder")	• Ta (as in "Taco")
• Ga (as in "Gum")	• Ti
• Ka (as in "Cup")	• Ti Ra Ki Ta
• Na (as in "Not")	• Ti Ta

Scat Sounds

• Ah and Uu for lower notes	• De	• Yah	• Yot
• Ee for higher notes	• Du	• Vah	• Shot
• Shwee	• Uu	• Dow	• Doot
• Skwee	• Shu	• Duh	• Dup
• Dwee	• Bu	• Bop	• Bup
• Bee	• Sku	• Dop	• Dut
• Vee	• Vu	• Vop	• Doot-n
• Zee	• Dah	• Bot	• Doodle-n
• Wee	• Sha	• Zot	• Dot-n
	• What	• Dit	• Dweedle-ee
	• Bah	• Dot	• Du-ee-ah

Appendix C: Forms

In the course of your journey, you might find that you need copyright forms to protect a composition. These forms include:

1   Form PA – For registering performing arts works (published or unpublished)
2   Short Form PA – For registering performing arts works where you are the sole author and copyright owner of the work, it is completely new, it wasn't created for hire, it is completely new, and it is *not* an audiovisual work.
3   Form SR – For sound recordings.

You can find all of these forms and information about how to complete and submit them at www.copyright.gov/register.

Appendix D: The Virtual Mixer™ Interface

Patent #5,812,688

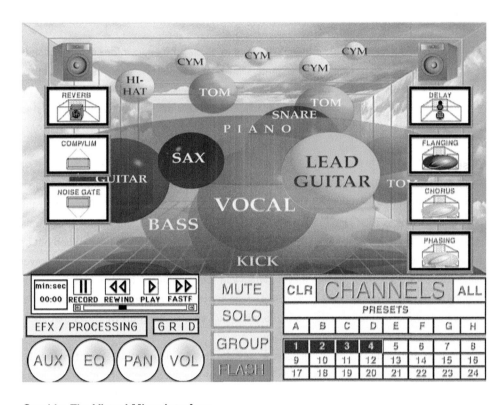

Graphic: The Virtual Mixer Interface

A 3D Visual Interface for Displaying Mixes and Controlling all Mixing Functions

Besides using this visual framework as an educational tool, it is now being developed as an interface for mixing consoles and effects. The interface uses three-dimensional representations of visuals of sounds between the speakers instead of two-dimensional pictures of a mixing board. It is intuitive and gives engineers (and students of engineering) more information about the mix, such as masking. The interface also works with 3D glasses for an enhanced visual presentation, and it is particularly useful for surround sound mixing.

Background

Over 20 years ago, we came up with the specific framework for mapping audio parameters into the visual world. The interface was originally developed as a tool to explain various styles of mixes, but then we realized it could be used to control a console. At that time, computers were not fast enough, and there were no MIDI-controllable consoles. The system is now being developed as an interface for digital consoles, and digital audio workstations.

What It Does

The Virtual Mixer is a graphical user interface that displays the amplitude and frequency range of sounds on a multitrack. It then allows you to use these visuals to control various types of mixing consoles and effect units.

The interface does two things:

1   It is used as a new type of "metering" system to display the audio information of sounds in a mix. Amplitude (same as the meters on a multitrack) is shown as a function of brightness, which makes the images flash to the music.
2   Movement of the visuals sends out MIDI information (or other automation protocols) to the console and effects to control all parameters.

Basics

Fader level (volume) is mapped out as a function of front to back so that louder sounds appear closer (and, therefore, a little larger), and softer sounds appear more distant. The grid on the floor is calibrated to the faders on the mixing console. Panning is naturally mapped out as a left-to-right placement. The average pitch of the sound over the entire song is mapped out as a function of up and down. Higher pitch sounds appear higher between the speakers; lower pitch sounds appear lower, just as they do when you focus on imaging. Also, high-frequency sounds are smaller because they take up less space in a mix. Low-frequency sounds are larger because they mask other sounds more in a mix. You can also bring up windows that show equalization and auxiliary send levels visually.

The next step is to put on 3D glasses to see the images floating in 3D right where the sounds seem to be between the speakers. In such a virtual environment, the mixer can use a dataglove to move the sounds around in the mix. If you want a sound louder, pull it toward you. If you want it in the left speaker, put it there. You can even toss sounds back and forth. With surround sound, the images are floating all around you – throughout the room. You can even bounce sounds off the walls. *Then with a Sound Chair, you can actually place sounds inside your body!!!*

Formats

The interface is being developed for a wide range of mixing equipment: digital consoles, sequencers, and digital audio workstations. We will also be putting out versions that will control various multi-effects units and will display specific synthesizer sounds as texturized spheres corresponding to the waveform.

Pricing

The basic interface for a digital console or workstation will be around $300 to $500. A touch-sensitive screen add-on will be around $500. A full 3D version (with 3D glasses on a computer monitor) would be under $1,000. Full virtual reality with head tracker display helmet will be around $1500. Full big-screen projection will be around $10,000. Actual prices will drop as demand for the interface increases and as 3D technology continues to come down in cost.

More Engrossing

When you see the mix visually, you are incorporating more of your senses into the musical process. The more senses you use (especially in a fleeting artistic endeavor such as mixing sound), the more your entire being is engulfed in the experience. And the more you are engrossed in all of the details, the easier it is to come up with a great mix.

More Intuitive

Pictures of sounds are one logical step closer to the music you are mixing than faders and knobs on a console. Manipulating visuals of the sounds themselves is much more intuitive than pictures of a console. Pictures of sounds are more like music than knobs.

More Conducive to the Creative Process

Studio equipment is notorious for getting in the way of the creative process. Faders and knobs on a console distract recording engineers from the music they are trying to mix. Images that flash to the music help the user to focus more clearly on the invisible sounds they are mixing.

More Information about the Mix

The visuals provide additional information helpful for the mix. The primary goal of the visual system has been to show "audio masking" visually. Engineers can use this information to discover hidden problems, and best of all, to be able to explain the problem to the band and producer. For example, a bad arrangement becomes clearly evident.

More Perspective

The visual framework provides the engineer with a perspective on all of the possibilities available to him. The framework shows each parameter within each piece of equipment in the studio and shows how each parameter contributes to an overall mix. With this visual perspective of all that is possible in a mix, a humongous number of possibilities are displayed in a way that puts an array of creative ideas at an engineer's fingertips.

More Relationships

The interface shows more than just all of the settings of each piece of equipment in the studio. The interface shows the *relationships* of all the settings and how they work together to create a mix. After all, it is the relationships of all the settings that really count.

Better Communication

The interface enhances and simplifies the communication process between recording engineer and client (band or producer). The interface is so intuitive that even inexperienced clients will be able to follow the development of the mix and communicate effectively with the engineer.

More Fun

And of course, we mustn't forget – flashing 3D visuals are a blast to watch and work with.

"The nature of the medium affects the art you create."

The Virtual Mixing Company
351 9th St. #202, San Francisco, CA 94103
David@GlobeRecording.com
415 777–2486
www.VirtualMixer.com
E-mail (or mail) us if you might be interested, and we'll let you know when it is available.

Glossary

ABA   The three-section form of a song that generally represents verse, chorus, and restatement of the verse.

Bass clef   Used for notes that are pitched in the lower register or range of an instrument and staff, such as bass guitar, orchestra, or upright bass, cello, and baritone. Also called the *F clef*.

Chords   Two or more notes played together at the same time.

Chorus   Sometimes called the *refrain*, it is the section of a song that is repeated after the verse. Also referred to as the *hook* or *B section*.

Circle of fifths   A chart that shows all the fifth intervals, which show you how many sharps and flats are included in the key. This system teaches you how to understand, approach, and play a song in any key.

Clef signs   Symbols that are written at the beginning of the staff and are used to indicate whether notes are to be played higher or lower on the piano (or any instrument) and placed higher or lower on the staff.

Dotted notes   When a dot is placed behind a note, it increases its duration by half its value.

Falls   A slurred downward sweep in pitch that falls off in volume. Normally done with horns.

First and second endings   Another repeat sign that indicates after section one and two are played, repeat section one, skip Part 2 and go to Part 3.

Flats   One half-step lower or to the left on the keyboard. When a flat sign is placed in front of a note on the staff, it lowers the note by a half-step.

Grand staff   The combination of both the treble and bass clefs using a brace and a line to connect the two.

Harmony   When you harmonize by playing two or more notes simultaneously, you can create an accompaniment with chords for the melody or melodic structure.

Hook   A catchy and memorable part in a song (this is how we use the term in this book). Another definition is that the hook is sometimes used to refer to the *chorus* of the song (particularly in the hip-hop community).

Improvise   To create solos from related scales within a song based on the key signature and chord structure of a piece of music executed during a particular passage or phrasing.

Intervals   Refers to the distance from one note to the next. For example, the interval distance between C and D is one whole step.

Inversions   A triad chord restructured, where the root is no longer in the first or bottom position.

Key signature   The sharps or flats located at the beginning of the music on the staff following the time signature indicate what key the song is in. If there are no sharps or flats, the song, in most cases, is in the key of C.

Lead-in   Same as *pickup*.

Ledger notes   When notes are too high or low to be written on the staff, ledger lines are used to extend the staff's range. Included is a line between the treble and bass clefs. This line indicates middle C. Ledger notes are placed on these floating ledger lines.

Major triad   A chord consisting of the root, major third, and major fifth of the scale. Also referred to as *tonic*, *maj third*, and *perfect fifth*.

Minor triad   A chord constructed from the root, flat third, or minor third and a perfect fifth.

Modulate   To change pitch or transpose to another key at a set point in a song.

Measures   The areas on the staff between the bar lines. Measures are sometimes referred to as *bars*.

Melody   Specifically refers to the main musical idea or a single line sung or played in a song; this idea is usually accompanied by chords in a song.

Note values   How long the note is to be played or held (also based on tempo). Values include whole, half, quarter, eighth, sixteenth, and so on.

Octave   The eighth interval of the major scale.

Phrase   A short part within a section of the music; a short musical idea executed vocally or instrumentally. Many times producers will use this term when referring to how many words are to be sung at a time within short musical sections of a song.

Pickup   Notes at the beginning of a song that make up an incomplete measure that usually come before the downbeat of a complete measure.

Podcast   Coined by Ben Hammersley and borrowed from Apple's iPod. Meaning an on-demand internet radio stream or broadcast. Podcasters are able to narrow-cast and are not restricted to ad revenue guidelines. Narrow-cast means a specific demographic unlike terrestrial radio (conventional) that has a much more broader audience.

Repeat signs   When two dots are placed before a double bar, this simply means to go back to the beginning and repeat the music again. Sometimes the first double bar lines appear with the two dots to the right of it. This indicates that this is the repeat point after you reach the repeat sign.

Rests Within music, silence or pause is indicated by musical notations known as *rest signs*. Their silence value is the equivalent of actual note values.

Scale A set pattern of notes with each note representing a degree or step in the sequence of that particular scale. Each scale has its own individual sound, character, and name.

Sharps One half-step higher or to the right on the keyboard. When a sharp sign is placed before a note on the staff, it increases the note by a half-step. *Sharp* is also a term used by musicians to indicate whether a note is slightly high and out of pitch. (This term is commonly used in the recording studio.) It is also a good beer to drink to keep your edge.

Shorthand charts Charts for reading music without staves. Normally these are chord charts.

Stab Loud, short duration notes. Normally with horns.

Staff Five lines and four spaces, with each representing a musical note sound or tone. Music on the staff is read from left to right.

Stream Another word for broadcast. Commonly music played live over the internet.

Syncopation When the accent in a musical phrase or passage accents or falls on the weak beat, not the strong beat.

Tempo How fast or slow a piece of music is to be played. Tempo is commonly measured in beats per minute. Sometimes this notation is written in European language.

Time signature Numbers that look like fractions that appear at the beginning of the music following the clef sign. The top number represents how many counts or beats per measure, and the bottom number states what kind of note receives the one beat.

Transpose To rewrite a song in its exact entirety in another key signature.

Treble clef Represents notes in and on higher pitch ranges from tenor to soprano and beyond, such as tenor, alto, and soprano voices, as well as flutes and other horns. Also called the *G clef* because it is a stylized G whose base encircles the G line on the staff.

Trill A quick back and forth between two short notes.

Verse The part of the song that establishes and tells the story. Also referred to as the *A section*.

VOD Casting Podcasting video as opposed to music.

Webserver A computer system that processes and distributes information on the World Wide Web.

WordPress Content management and blog publishing application.

About the Authors

David Gibson has produced albums in a wide range of styles of music – acoustic rock, jazz, rock, new age, heavy metal, rap, hip-hop, and electronica. Gibson has been teaching, engineering, and producing groups in major 24-track studios since 1982 and was the founder and owner of Globe Institute in San Francisco. Before that, he taught recording at Cabrillo College in Santa Cruz, California, for five years. He has been using the concepts and visuals presented in this book in his classes since 1986.

Previously, Gibson owned a commercial music production company called Creative Music Services in Atlanta, Georgia. He has recorded many jingles for the Santa Cruz Beach Boardwalk.

David has also done recording for James Brown's band, Bobby Whitlock (Derek and the Dominos), the Atlanta Rhythm Section, and Hank Williams Jr.'s band, as well as having worked with members of the Doobie Brothers, Lacy J. Dalton's band, Herbie Hancock's band, and It's a Beautiful Day.

Gibson is also the inventor of the patented Virtual Mixer, which displays a mix visually in real time. He created a customized version that shows the visuals in 3D with 3D glasses.

Gibson is the author of *The Art of Mixing*. He is also the producer and director of the video series *The Art of Mixing*. Since writing the first version of this book David has now become one of the leaders in the field of Sound Healing and Therapy.

David Gibson is the founder and director of the Globe Sound and Consciousness Institute in San Francisco, offering Individual Classes and State approved Certificate and Associates Degree Programs in Sound Healing and Therapy (www.SoundHealingCenter.com). The Institute also offers a Certificate Program in Recording Arts and Technology. All programs are also available online with live instructors or by video.

David is the author of the #1 selling book in the field of Sound Healing, "The Complete Guide to Sound Healing" (www.CompleteGuidetoSoundHealing). He is also a top selling producer of Sound Healing music (www.SoundHealingCenter.com/store), and runs the Sound Therapy Center at the Institute (and online), offering over 15 types of sound healing treatments (www.SoundTherapyCenter.com). David is the founder of the Sound Healing Research Foundation (www.SoundHealingResearchFoundation.org) designed to help bring sound healing into the mainstream (homes & hospitals).

David has a unique understanding of how sound and music create deep and lasting effects on the mind, body, and spirit. Having run the largest Sound Healing Institute in the nation for over 15 years (with 25 instructors), and having put on seven International Sound Healing

Conferences, David has developed a complete perspective of everything going on in the field – particularly how sound affects us physically, mentally, emotionally, and spiritually. David's education in Physics at the University of California at Berkeley gives him a unique basis for understanding and explaining the way that sound affects us physically.

David has been exploring the use of binaural beats for brainwave entrainment for over 15 years and has over a hundred CDs that incorporate brain enhancing frequencies in the music.

The CD catalog includes songs for relaxation, sleep, depression, anxiety, enlightenment, opening the heart, connecting to Spirit, mental clarity, sexual energy, overcoming fear, releasing grief, harmonizing emotions and balancing chakras. David did part of the soundtrack for the movie, "The Living Matrix," and two of his CDs have quantum healing fields embedded in the music. The embedded frequencies are based on years of research by Nutri-Energetics on the resonant frequencies of morphogenetic fields. David also has seven CDs for surgery (pre-op, during surgery, and post-op) as well as guided meditations for relaxation, releasing anxiety and pain, releasing stuck emotions, and bringing unconditional love into the heart. Most of the CDs in the catalog incorporate binaural beats to entrain the listener into even deeper states.

In order to bring sound healing into the mainstream, David has set up the Sound Healing Research Foundation to find and study all of the frequencies and musical intervals within the body. He has completed a research project on using sound and music to help Parkinson's patients. The research foundation is working on projects for using sound for sleep enhancement, pain management, ADD/ADHD, PTSD, and autism.

The Research Foundation also received a large grant to take the Institute's Sound Healing curriculum (including Music Theory and Geometry) into two Montessori schools. Once completely implemented, an association will be set up to bring this curriculum into public and private schools around the world.

David also runs the Sound Healing Metaphysical store – offering a full range of sound healing instruments and vibrational tools (www.SoundOfLove.com).

You can make contact via: David@GlobeRecording.com (Feel free to e-mail with comments or questions.)

Maestro Curtis is a seasoned musician, (pianist/multi-instrumentalist) pianist, vocalist, arranger, producer, composer, sound alchemist, community activist and educator. He is an all-round entertainer whose musical impact has been felt in the U.S. and the world over. His piano performances have been compared to legendary pianists such as Joe Sample, Thelonious Monk, and Bud Powell; his song writing is lyrically poetic, melodically memorable, and his arrangements are uniquely his own. The world-renowned pianist, Ricardo Scales, calls Maestro, "the man with the golden voice and the magic hands".

The artists he has written and produced music for are astounded at his ability to capture their unique and individual point of view and enhance it. Maestro's singing has been likened to Donny Hathaway, Al Jarreau, Rance Allen, and Peabo Bryson. He is able to go in and out of genres with a seamless ease, both vocally and musically.

Maestro takes pride in his work as a college professor: he taught in the San Francisco Unified School District, Houston Independent School District, City College of San Francisco, San Francisco State University, San Francisco Library outreach programs and is proud to be a part of the faculty at the Community Music Center, the oldest music institution in the country, founded by Gertrude Field, where he teaches piano, voice, guitar, music theory, orchestration, and arrangement.

Maestro, as a recording artist, has been signed to the following labels: Kalimba Records, MCA, Sony, Silhouette, Castle Rock. Maestro is the founder of his own record label JazzyBoo Records.

Maestro is a protégé of the late great Maurice White (leader and founder of Earth, Wind & Fire, who called Maestro a musical genius), has composed and produced music with the legendary Hubert Eaves III, D Train (who has written and produced artists such as Whitney Houston and Stephanie Mills, to name a few), who is the father of the "Minneapolis Sound" who influenced the likes of Prince, Terry Lewis and Jimmy Jam. Maestro has performed with Lenny Williams, studied under and shared the stage with Frank Foster, Larry Gales, and Ben Riley, the members of "Sphere", the Count Basie Orchestra and Duke Ellington Orchestra, to name a few. Maestro's list of notables is impressive, however, the following associated acts such as the original lead singer of the gospel smash hit, "Oh Happy Day," he produced music for Dorothy Morrison, Papa Pete Escovedo and his family known as the "The E Family," he is happy to refer to as family. It was also Maestro's pleasure to perform with Grammy winning disco icon, Thelma Houston, and legendary queen of the Castro scene, Sylvester. In the gospel world there are few names that are bigger than that of James Cleveland, the Prince of Gospel, and Maestro has performed and toured with him along with many of the famed and celebrated gospel singers who toured in that same camp, such as Daryl Coley, Billy Preston, Glenn Jones, and Keith Pringle.

Under the direction of Dr. Frank Foster, Maestro played with the Count Basie Orchestra, while he was a college student at the historic black college, Grambling State University. During his undergrad and graduate studies at Grambling, Maestro also played in the Duke Ellington Orchestra, under the direction of Paul and Mercer Ellington.

Maestro Curtis and his group, Xpression, was signed to Maurice White's (founder and lead vocalist of Earth, Wind & Fire) record label, Kalimba Records, as a producer, vocalist, pianist, and guitarist in 2000. The group's debut album, *Power*, was released in the early part of 2000. Maestro wrote, produced, and arranged all of the songs on the album, except for the three songs that were either written or co-written with Paul Laurence, who had worked with Grammy winning R&B vocalist Freddie Jackson.

In the liner notes of Xpression's album, *Power*, Maurice White stated:

> The band is named Xpression because when all five members of the band come together to play it is a true expression of their musical cohesion. The music on this record is the result of the collaboration between very accomplished musicians, it is rooted in classic R&B with jazz undercurrents tailored for commercial appeal. It covers all the bases, it's energetic, sincere and sensitive. I hope you like it as much as I have enjoyed helping with its creation.

Maestro sits on the board of directors and is the vice president of artist development for Hitman Records, an independent record label whose roster of artists span nationwide. The CEO of Hitman Records and Maestro were in an article "Topping The Charts" in the nationally circulated magazine, *Black Enterprise*, where he is pictured alongside C. Michael Brae, CEO of

Hitman Records, working in the recording studio. The article describes how to navigate the music industry, utilizing techniques that successful independent record labels use to develop and establish new talent.

Maestro shares his knowledge of the music business in and out of his own private teaching sessions, via workshops and music business classes. Maestro serves as a panelist at several conferences such as Success in the Bay "What You Better Know About the Music Business" Seminar, California Lawyers for the Arts, 2008 Music Business Seminar, and in June 2010 he appeared at the Soul Music By the Bay Music Conference, where he spoke of the importance of studying music, empowering yourself as an artist, sharing his knowledge of the music industry, as well as sharing his own personal experiences as an entertainer, musician, and producer from when he was signed to major record labels, independent labels, and owning his own record company.

Maestro created a fraternity of musicians, the Jazz Hieroglyphics, where membership has grown to exceed 60 over the span of 30 years, members who span the globe. Maestro heads several ensembles that can be as small as a quartet and can be as large as a 22-piece big band. One of his featured ensembles, "The S.O.L. (Spirit of Love) Funkestra," with his wife Nola Curtis as the lead singer, is in its genesis, gearing up for a world tour, pushing their latest single, "Power (Power to the People)," which also features Larry Douglas, former music director for Johnny and Shuggie Otis.

Maestro and the choir of San Francisco's historical Macedonia Missionary Baptist Church starred in Ron Howard's, *Parenthood* (2010 TV Series), on the pilot episode, where Maestro's voice is featured ushering in the comical "Frozen Sperm" discussion between the series characters, Crosby (Dax Shepard) and Katie (Marguerite Moreau). Maestro also starred in an independent Science Fiction/Action Film, *Origin*, where he plays dual roles, which is being released in episodes on YouTube.

Maestro serves as the musical director for Darcel Walker's writer/director, sci fi series, *Starlight Source*, ever expanding his musical reach into television and film.

Maestro Curtis is an integral figure of the Sound Healing Community in North America. He worked with United Way to develop the first comprehensive music therapy program in the world for people suffering from Azheimer's and Dementia. In addition to his private sound consciousness practice, he also taught at one of the foremost sound healing institutes in the world, The Globe Sound and Consciousness Institute located in San Francisco. In 2007, the World Sound Healing Conference held in San Francisco, CA, he was one of the keynote speakers and musicians. During the final concert of the conference Maestro performed with legendary guitarist Stanley Jordan.

Maestro played on David Gibson's project,"Body Field Sound Healing." In this project the music is imprinted with Quantum Fields expressed frequencies, notes, and chord progressions. It is referred to as "Imprinted Music," and is encoded with information to enhance your well-being. Maestro infuses his music with sound healing techniques.

Maestro has over 40 years of community service experience, which he continues today through his work with Older Adult Choirs, which are free to seniors who reside in the San Francisco/Bay Area, in partnership with the University of California San Francisco, Community Music Center, Bayview Hunter Point Senior Services, and I.T. Bookman Community Center. Along with his wife he has led them to perform in venues like San Francisco City Hall, Herbst

Theatre, and the older adult choirs' voices and images are featured in an immersive art installation and on-going research and interview project called "Worth Your Salt" by San Francisco artist, Annie Albagli, where Maestro and his sound healing techniques and philosophy are featured in his live workshops in the opening celebration near City Hall in the "Please Touch Community Garden" (offered for free to serve the community at large). Maestro and Nola's work with their Older Adult Choirs was featured on *USA Today*, along with the creator of the study that gave birth to the Older Adult Choir Program, Julene Johnson, a cognitive neuroscientist and professor at UC San Francisco School of Nursing's Institute for Health & Aging.

Maestro's vast musical experience more than qualifies him to teach at any institution of his choosing, however, he finds great joy at serving his community and teaching at the San Francisco Community Music Center where students are offered income-based tuition, therefore Maestro can serve a far larger number of students. Maestro emphasizes music's power to reinforce, strengthen critical thinking, and feed the soul. One of his favorite sayings is that "music is everyone's birthright," along with "we are music, the instrument is the tool to express yourself."

Finally, the education that is the foundation of Maestro's philosophy, performance, and teaching style is listed here: three Bachelor's Degrees: 1) Piano and Vocal Music, 2) Radio, Television, and Film Production, and 3) Speech and Theatre, two Masters Degrees: 1) Music Theory/Orchestration and Arranging, and 2) Philosophy, and his PhD: Human and Organizational Development/Symbolic Analysis with an emphasis on Educational Leadership and Change (The Fielding Institute in Santa Barbara, California).

Index

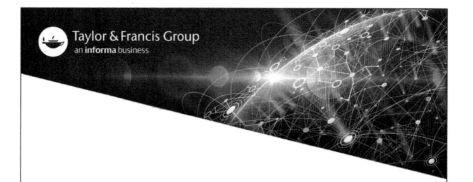

9780815369387